Lecture Notes in Mathematics

Edited by A. Dold and B. Eckmann

679

Numerical Treatment of Differential Equations in Applications

Proceedings, Oberwolfach, Germany, December 1977

Edited by
R. Ansorge and W. Törnig

Springer-Verlag
Berlin Heidelberg New York 1978

Editors
Rainer Ansorge
Institut für Angewandte Mathematik
Universität Hamburg
Bundesstraße 55
D-2000 Hamburg

Willi Törnig
Fachbereich Mathematik
Technische Hochschule Darmstadt
Schloßgartenstraße 7
D-6100 Darmstadt

Library of Congress Cataloging in Publication Data
Main entry under title:

Numerical treatment of differential equations in applications.

(Lecture notes in mathematics ; 679)
"Meeting on numerical treatment of differential equations, held in the Mathematical Research Center of Oberwolfach."
Bibliography: p.
Includes index.
1. Differential equations--Numerical solutions--Congresses. I. Ansorge, Rainer. II. Törnig, W. III. Series: Lecture notes in mathematics (Berlin) ; 679.
QA3.L28 no. 679 [QA370] 510'.8s [519.4]
78-11883

AMS Subject Classifications (1970): 34C15, 65-02, 65Lxx, 65Mxx, 65Nxx

ISBN 3-540-08940-3 Springer-Verlag Berlin Heidelberg New York
ISBN 0-387-08940-3 Springer-Verlag New York Heidelberg Berlin

Printing and binding: Beltz Offsetdruck, Hemsbach/Bergstr.
2141/3140-543210

Foreword

The meeting on numerical treatment of differential equations, held in the Mathematical Research Center of Oberwolfach, West-Germany (December, 12-16, 1977), was attended by mathematicians as well as by engineers, scientists, and economists.

One of the aims of the conference was to foster cooperation between representatives of these fields, at least with respect to the topic of the meeting.

It was very interesting for the attending mathematicians to become acquainted with new mathematical problems (and some methods to solve them), arising e.g. from engineering, which are unconventional and which therefore have not yet been treated by mathematicians.

On the other hand, the participating non-mathematicians took advantage of the opportunity to inform themselves intensively about new methods and results in numerical analysis.

Many new ideas were presented at the conference; a great part of them now appear in these notes.

We want to thank the director of the Oberwolfach-Institut, Prof. Barner, who gave us the opportunity for organizing this meeting.

We also pay tribute to Dr. Kreth, Hamburg, who coordinated the production of the copy-ready manuscript.

Last but not least we thank the editors of the Lecture Notes series and the Springer-Verlag for the speedy publishing of this volume.

Hamburg and Darmstadt, Mai 1978

R. Ansorge W. Törnig

Contents

List of Contributors

Bohl, E. Prof. Dr.
Fachbereich Mathematik der
Universität Münster
Roxeler Straße 64
D - 4400 Münster

Mc Cormick, S.F.
Department of Mathematics
Colorado State University
Fort Collins
Colorado 80 523/USA

Elben, W. Dipl. Math.
AEG-Software-Center
Goldsteinstraße 235
D - 6000 Frankfurt-Niederrad

Koßmann, H. Dr.
Ruhr University, Math. Inst. NA
D - 4630 Bochum

Kreth, H. Dr.
Institut für Angewandte Mathematik
der Universität Hamburg
Bundesstraße 55
D - 2000 Hamburg 13

Lambert, J.D. Prof. Dr.
Department of Mathematics
University of Dundee
Scotland

Mehri, B. Prof. Dr.
Aria-Mehr-University
Department of Mathematics
Tehran
Iran

Miller, J.J.H. Prof. Dr.
Mathematisch Instituut
Katholieke Universiteit
Nijmegen
The Netherlands

Müller, W. Prof. Dr.
Technische Hochschule Darmstadt
Fachbereich 17
Institut für elektrische Energiewandlung
D - 6100 Darmstadt

Nixdorff, K. Prof. Dr.
Hochschule der Bundeswehr Hamburg
Fachbereich Maschinenbau
Holstenhofweg 85
D - 2000 Hamburg 70

Nørsett, S.P. Prof. Dr.
Norges Tekniske Høgskule
Instituut for Numerisk Mathematikk
N - 7034 Trondheim

Ortiz, E.L. Prof. Dr.
Imperial College
University of London
London
England

Rautmann, R. Prof. Dr.
Fachbereich Mathematik-Informatik
Gesamthochschule Paderborn
D - 4790 Paderborn

Wolff, W. Dipl. Phys.
AEG-Software-Center
Goldsteinstraße 235
D - 6000 Frankfurt-Niederrad

Wirtz, W. Dr.
Route du Lion 172
B - 1420 Braine L'Alleud

On two boundary value problems in nonlinear elasticity from a numerical viewpoint

Erich Bohl

This paper is concerned with boundary value problems of the general form

BVP: $$-x''-\alpha t^{-1}x' = f(t,x,\lambda) \quad \text{on} \quad [0,1]$$
$$\alpha_0 x(0)-\beta_0 x'(0) = \alpha_1 x(1)+\beta_1 x'(1) = 0$$

with a real parameter $\lambda > 0$ and reals $\alpha, \alpha_i, \beta_i$ satisfying

$$\alpha_i \geq 0, \ \beta_i \geq 0, \ \alpha_i+\beta_i > 0 \quad (i=0,1).$$

In section 1 we consider BVP under assumptions on f which are satisfied for the circular membrane problem recently discussed in [8]. In this paper H. Weinitschke treats the problem using Schauder s fixed point theorem. In contrast to this our analysis is based on the more elementary contraction mapping principle. However, our iterative procedure converging to the solution is different from the one in [8]. H. Weinitschke notes that his iterative method does not come from a contraction for all positive values of λ. We add a stability inequality (Theorem 1.1) yielding uniqueness of the solution as well as error estimates for it and we conclude with a numerical example.

In section 2 we change the restrictions on f to cover the problem of the rotating string [4b,7]. Here the emphasis is on the discussion of a finite difference analogue to the continuous problem. This is a finite dimensional bifurcation problem. It is shown that the branch of nonnegative solutions bifurcating at the smallest positive eigenvalue of the corresponding linearization may easily be computed by standard direct iterative procedures. No imbedding technique is needed and hence no information on the solution at a parameter $\lambda \neq \bar{\lambda}$ if the solution for $\lambda = \bar{\lambda}$ is wanted. To illustrate the method we give some figures for the rotating string problem.

0. Basic notations

Our setting is a vector space X of bounded real valued functions on a nonempty set Ω. X is endowed with the sup-norm $\| \|_{\delta}$ and the partial ordering

$$x \leq y \Leftrightarrow x(t) \leq y(t) \text{ for all } t \in \Omega$$

with the underlying cone $X_+ = \{x \in X : x \geq \theta \text{ (=zero element)}\}$, cf. [2a]. We shall use order intervals $[x,y] = \{x \in X : x \leq z \leq y\}$ and the notation $|x| \in X$ for the "absolut value" of $x \in X$ defined via

$$(1) \qquad |x|(t) = |x(t)| \text{ for all } t \in \Omega.$$

L[X] denotes the set of all bounded linear operators on X and $L_+[X]$ stands for the set of all monotone elements of L[X]. An operator T taking a subset V of X into X is called monotone if

$$x \leq y \Rightarrow Tx \leq Ty$$

for all $x,y \in V$. Let T,S be operators from $V \subset X$ into X then we write $S \leq T$ if T-S is monotone.

Let V be a linear subset of X then we put $V_+ = V \cap X_+$. A linear operator A from V into X is called inverse-monotone (for short i.m.) if the inverse A^{-1} from X into V exists and if $A^{-1} \in L_+[X]$.

The setting described so far is being applied in a "continuous" and in a "discrete" situation:

The "continuous" version puts

$$(2) \qquad \begin{aligned} \Omega &= [0,1], \ X = C[0,1] =: C \\ V &= \{x \in C^2[0,1] : \alpha_0 x(0) - \beta_0 x'(0) = \alpha_1 x(1) + \beta_1 x'(1) = 0\}. \end{aligned}$$

The "discrete" version defines the grid points $t_j = jh$, $j = 0,..,M$ in [0,1] for $h = M^{-1}$, $M \in \mathbb{N}$ and then puts

$$\Omega = \Omega_h = \{t_j : j = 0,..,M\}, \ X = \mathbb{R}^{\Omega_h}.$$

Here, we use the notation

$$L^h := L[\mathbb{R}^{\Omega_h}], \quad L_+^h := L_+[\mathbb{R}^{\Omega_h}].$$

For $x \in C$ we denote by $x_h \in \mathbb{R}^{\Omega_h}$ the restriction of x to the grid Ω_h, e.g. the function $\delta(t) = 1$ on $[0,1]$ defines the vector $\delta_h \in \mathbb{R}^{\Omega_h}$ with components $\delta_h(t) = 1$ for $t \in \Omega_h$.

1. The membrane problem

In [8] H. Weinitschke considers the differential equation

(3a)
$$-x'' - 3t^{-1}x' = 4\{h(t,x,\lambda)\ (x+\lambda+h(t,x,\lambda))\}^{-1}$$
$$h(t,x,\lambda) = \sqrt{(x+\lambda)^2+k^2t^2}, \quad k \in \mathbb{R}, \ \lambda > 0$$

subject to one of the boundary conditions

(3b) $\quad x'(0) = x(1) = 0 \quad \text{or} \quad x'(0) = ax(1)+x'(1) = 0 \quad (a > 0).$

The problem (3a,b) occurs in the study of a circular membrane of radius 1 under vertical pressure (cf. [4b,8] and the references given therein). It is a special case of the boundary value problem

(4) $\quad -x''-3t^{-1}x' = f(t,x,\lambda)$ on $[0,1]$, $x'(0) = \alpha x(1)+\beta x'(1) = 0$

if we assume

A1: $\alpha > 0,\ \beta \geq 0$;
F1: $f,\ D_2f := f_x \in C([0,1]\times\mathbb{R}_+^2)$;
F2: $f(t,v,\lambda) \geq 0,\ D_2f(t,v,\lambda) \leq 0$ on $[0,1]\times\mathbb{R}_+,\ \lambda > 0$;

where $\mathbb{R}_+$ denotes the set of all nonnegative reals.

To study (4) let us consider the differential operator

$$L:\ x \longmapsto -x''-3t^{-1}x'-r(t)x \qquad (r \in C)$$

from V into C (see (2)).

The equation

$$(5) \qquad Lx = g \qquad (g \in C)$$

is equivalent to

$$x(t)-\{\int_t^1 \int_0^s (\tau s^{-1})^3 r(\tau)x(\tau)d\tau ds+\beta\alpha^{-1} \int_0^1 \tau^3 r(\tau)x(\tau)d\tau\}$$
$$= \int_t^1 \int_0^s (\tau s^{-1})^3 g(\tau)d\tau ds+\beta\alpha^{-1} \int_0^1 \tau^3 g(\tau)d\tau.$$

This is a Fredholm equation in the Banach space $(C, \| \|_\delta)$ with a completely continuous integral operator. Hence, Fredholm's alternative holds for (5). Furthermore, for any $x \in V$ such that $Lx \geq \theta$ we have $x \geq \theta$ if we assume $r \leq \theta$ [6, Chap. 1, Th. 3]. This establishes our

<u>Lemma</u> 1: Let $r \in C$, $r(t) \leq 0$ on $[0,1]$. Then the operator L: $V \longrightarrow C$ has a monotone inverse $L^{-1}: C \longrightarrow V$.

Next, we define the operators

$$(6) \qquad A: x \longrightarrow -x'' - 3t^{-1}x', \quad F_\lambda : x \longrightarrow f(\cdot, x, \lambda)$$

mapping V into C, C into C, respectively.

<u>Lemma</u> 2: Let $r \in C$, $r(t) \leq 0$ on $[0,1]$. Then for any $x \in C$, $x \geq \theta$ we have $L^{-1}x \leq A^{-1}x$.

<u>Proof</u>: $x \geq \theta$ implies $A^{-1}x \geq \theta$ and hence $x = A(A^{-1}x) \leq L(A^{-1}x)$ since $r \leq \theta$. An application of L^{-1} completes the proof using Lemma 1.

Let $x,y \in V_+ = V \cap C_+$ (cf. section 0). Then

$$(7) \qquad (A-F_\lambda)x-(A-F_\lambda)y = A(x-y)-r(\cdot)\,(x-y) =: L(x-y)$$
$$r(t) = \int_0^1 D_2 f(t,y(t)+\tau(x(t)-y(t)))d\tau.$$

By F2 we have $r(t) \leq 0$ on $[0,1]$, hence using Lemma 1 and Lemma 2 we find

(8) $\quad |x-y| = |L^{-1}(L(x-y))| \leq L^{-1}|L(x-y)| \leq A^{-1}|L(x-y)|,$

where we have adopted the notation introduced by (1). The formulae (7) and (8) yield our

Theorem 1.1: Let $\lambda > 0$ and let A1, F1 and F2 be satisfied. Then the operator $A: V \longrightarrow C$ defined by (6) has a monotone inverse $A^{-1}: C \longrightarrow V$ and for any two functions $x,y \in V_+$ we have the stability inequality

(9) $$|x-y| \leq A^{-1}|(A-F_\lambda)x-(A-F_\lambda)y|,$$

where $F_\lambda: C \longrightarrow C$ is defined in (6).

Next we consider the nonlinear problem

(10) $$Ax = F_\lambda x$$

with A and F_λ as given in Theorem 1.1. By Theorem 1.1 there is at most one solution $\bar{x}_\lambda \in V_+$ of (10) and if this solution exists then by (9) we have the error estimate

$$|\bar{x}_\lambda - y| \leq A^{-1}|(A-F_\lambda)y| \quad \text{for any } y \in V_+.$$

In particular $y = \theta$ yields

$$\theta \leq \bar{x}_\lambda \leq A^{-1}F_\lambda\theta$$

(note that $\bar{x}_\lambda \in V_+$ and that $|F_\lambda\theta| = F_\lambda\theta$ by F2).

Let us prove that $\bar{x}_\lambda$ indeed exists. To this end we choose $N \in \mathbb{N}$ such that

$$A^{-1}F_\lambda\theta \leq N\delta.$$

Then we define the function

$$f^N(t,v,\lambda) = \begin{cases} D_2f(t,0,\lambda)v+f(t,0,\lambda) & v < 0 \\ f(t,v,\lambda) & 0 \le v \le N \\ D_2f(t,N,\lambda)\ (v-N)+f(t,N,\lambda) & N < v \end{cases}$$

for $t \in [0,1]$, $\lambda > 0$. This function satisfies $D_2f^N(t,v,\lambda) \le 0$ on $[0,1]\times\mathbb{R}$. Therefore the iterative procedure

$$(11) \qquad (A-S_\lambda)x^{n+1} = (F_\lambda^N - S_\lambda)x^n \qquad (n \in \mathbb{N})$$

converges for any $x^0 \in C$ to the solution $\bar{x}_\lambda^N \in V$ of

$$Ax = F_\lambda^N x$$

(see [1]). Here F_λ^N is defined via

$$F_\lambda^N\colon\ x \longrightarrow f^N(\cdot,x,\lambda)$$

and S_λ is any operator on C of the form

$$(12a) \qquad S_\lambda x = s_\lambda(\cdot)x$$

with a function $s_\lambda \in C$ such that

$$(12b) \qquad 2s_\lambda(t) \le D_2f(t,v,\lambda) \text{ on } [0,1]\times[0,N].$$

Obviously, Theorem 1.1 applies to the operator $A-F_\lambda^N$ and since $D_2f^N(t,v,\lambda) \le 0$ holds globally on $[0,1]\times\mathbb{R}$ the stability inequality (9) is true for all $x,y \in V$ if we replace F_λ by F_λ^N (see the proof of Theorem 1.1). In particular we have

$$(13) \qquad |\bar{x}_\lambda^N| \le A^{-1}F_\lambda^N\theta = A^{-1}F_\lambda\theta \le N\delta$$

or $\bar{x}_\lambda^N(t) \le N$ on $[0,1]$. This implies $f^N(t,\bar{x}_\lambda^N(t),\lambda) \ge 0$ on $[0,1]$ by the definition of f^N and hence $F_\lambda^N\bar{x}_\lambda^N \ge \theta$ or $\bar{x}_\lambda^N = A^{-1}F_\lambda^N\bar{x}_\lambda^N \ge \theta$. This together with (13) shows

$$\theta \leq \bar{x}_\lambda^N \leq N\delta,$$

hence $A\bar{x}_\lambda^N = F_\lambda^N \bar{x}_\lambda^N = F_\lambda \bar{x}_\lambda$, which proves our

Theorem 1.2: Let $\lambda > 0$ and let A1, F1 and F2 be satisfied. The equation (10) has a unique solution $\bar{x}_\lambda \in V_+$. This solution is the limit of the globally convergent iterative process (11) where S_λ is defined via (12a,b).

Remark: H. Weinitschke [8] constructs $\bar{x}_\lambda$ using the iterative procedure

$$(14) \qquad Ay^{n+1} = F_\lambda y^n, \; y^0 = \theta.$$

Then $y^1 = A^{-1}F_\lambda\theta$ and $[\theta, y^1]$ is invariant under $A^{-1}F_\lambda$, so that Schauder's theorem guarantees a solution of (10) in $[\theta, y^1]$. In contrust to this way of reasoning in Weinitschke's paper the convergence of the process (11) is based on the contraction mapping Theorem (see [1]). As Weinitschke notes $A^{-1}F_\lambda$ is not for all $\lambda > 0$ a contraction and he refers for that to [4a]. Hence, the contraction mapping theorem does not apply to the process (14) in general.

2. The rotating string problem

The motion of a string of unit length fixed at one end and free to rotate at the other end is completely described if the solutions of the boundary value problem [4b,7]

$$(15) \qquad -x'' = \lambda x(t^2+x^2)^{-1/2}, \; x(0) = x'(1) = 0$$

are known. Clearly, this is a special case of

$$(16) \qquad -x'' = \lambda g(t,x), \; x(0) = x'(1) = 0$$

if we assume

G 1: $g(t,\cdot)$, $D_2g(t,\cdot) \in C(\mathbb{R}_+)$ for any (fixed) $t \in [0,1]$;

G 2: $D_2g(t,\cdot)$ is nonnegative and monotone decreasing on some interval $[0,w(t)]$ $(0 \leq w(t))$ for any (fixed) $t \in [0,1]$.

Note that we may take $w(t) \geq 0$ arbitrary for the rotating string problem.

Let us consider the finite difference analogue of (16) given by the set of equations

$$x(0) = 0$$

$$h^{-2}(-x(0)+2x(h)-x(2h)) = \lambda g(h,x(h))$$

$$\frac{h^{-2}}{12}(x(t-2h)-16x(t-h)+30x(t)-16x(t+h)+x(t+2h))$$
$$= \lambda g(t,x(t)) \quad (t=2h,..,1-2h)$$

$$h^{-2}(x(1-3h)-15x(1-2h)+27x(1-h)-13x(1))$$
$$= 11\lambda g(1-h,x(1-h))+\lambda g(1,x(1))$$

$$h^{-2}(-8x(1-3h)+54x(1-2h)-216x(1-h)+170x(1))$$
$$= 36\lambda g(1,x(1)).$$

The underlying grid Ω_h is given in section 0. Let G_h be the mapping on $\mathbb{R}^{\Omega_h}$ which assigns to $x \in \mathbb{R}^{\Omega_h}$ the vector whose t-th componer is

(17) $$(G_h x)\ (t) = g(t,x(t)) \qquad (t \in \Omega_h).$$

Our system is of the form

(18) $$A_h x = \lambda B_h G_h x$$

where $A_h \in L^h$, $B_h \in L^h_+$ are easily recognized from the explicit representation of the system. In all what follows we are merely concerned the set of nonlinear equations (18) where $h^{-1} = M \geq 4$.

In [3,5] it is shown that A_h is inverse monotone (i.m.), i.e. A_h^{-1} exists and belongs to L^h_+. By G1, G2 the matrix $DG_h(x)$ exists for all $x \in \mathbb{R}_+^{\Omega_h} = (\mathbb{R}^{\Omega_h})_+$ and $DG_h(\theta) \in L^h_+$. Since $B_h \in L^h_+$ we have $A_h^{-1}B_hDG_h(\theta) \in L^h_+$. Hence, there exists the smallest positive eigenvalue $\lambda_h > 0$ of the eigenvalue problem

$$A_h x = \mu B_h DG_h(\theta) x$$

and $A_h-\lambda B_h DG_h(\theta)$ is i.m. for $0<\lambda<\lambda_h$ (see [2a]).

<u>Theorem</u> 2.1: Let $4h \le 1$, $0<\lambda<\lambda_h$ and let G1 and G2 be satisfied. Then $A_h-\lambda B_h DG_h(\theta)$ is i.m. and the stability inequality

$$(19)\qquad |x-y| \le (A_h-\lambda B_h DG_h(\theta))^{-1}|(A_h-\lambda B_h G_h)x-(A_h-\lambda B_h G_h)y|$$

holds for any $x,y\in[\theta,w_h]$ where w_h is again the restriction of w from G2 to the grid Ω_h.

In particular, there is at most one solution of (18) in $[\theta,w_h]$. For the discrete rotating string problem any $w(t)\ge 0$ satisfies G2 and $B_h G_h\theta=\theta$. Hence, if $0<\lambda<\lambda_h$ its only solution in $\mathbb{R}_+^{\Omega_h}$ is $\bar{x}_\lambda=\theta$.

<u>Proof of Theorem</u> 2.1: Let $x,y\in[\theta,w_h]$. Then there exists $\eta_h\in[\theta,w_h]$ such that

$$(20)\qquad (A_h-\lambda B_h G_h)x-(A_h-\lambda B_h G_h)y=(A_h-\lambda B_h DG_h(\eta_h))\,(x-y).$$

From $A_h-\lambda B_h DG_h(\theta)\le A_h-\lambda B_h DG_h(\eta_h)\le A_h$ (use G2!) together with the inverse-monotonicity of $A_h-\lambda B_h DG_h(\theta)$ and A_h we may conclude that $A_h-\lambda B_h DG_h(\eta_h)$ is i.m. and that $(A_h-\lambda B_h DG_h(\eta_h))^{-1}\le(A_h-\lambda B_h DG_h(\theta))^{-1}$. This shows

$$|x-y| = |(A_h-\lambda B_h DG_h(\eta_h))^{-1}(A_h-\lambda B_h DG_h(\eta_h))\,(x-y)|$$
$$\le (A_h-\lambda B_h DG_h(\theta))^{-1}|(A_h-\lambda B_h DG_h(\eta_h))\,(x-y)|$$

and completes the proof if we apply (20).

To study the more interesting situation $\lambda_h<\lambda$ we will have to assume apart from G1, G2 also

G3: There exist $v\in[\theta,w_h]$, $\sigma\in(0,\lambda_h)$ such that

$$\lambda D_2 g(t,v(t)) = \sigma D_2 g(t,0)\qquad \text{for } t\in\Omega_h$$

Note that this is satisfied with $\sigma = \lambda$, $v = \theta$ if $0 < \lambda < \lambda_h$. For the rotating string G3 holds for any $\lambda > \lambda_h$: just pick $\sigma \in (0,\lambda_h)$ arbitrary and choose

(21) $$v(t) = \{(\lambda\sigma^{-1})^{2/3}-1\}^{1/2} t \qquad (t \in \Omega_h)$$

Based on G3 we construct the function

$$g^v(t,s) = \begin{cases} g(t,v(t)) & s < v(t) \\ g(t,s) & v(t) \leq s \leq w(t) \\ D_2 g(t,w(t))\,(s-w(t))+g(t,w(t)) & w(t) < s \end{cases}$$

$t \in \Omega_h$. For any $t \in \Omega_h$ we have

$$0 \leq g^v(t,x)-g^v(t,y) \leq D_2 g(t,v(t))\,(x-y) \quad \text{for } y \leq x, x,y \in \mathbb{R}$$

Therefore [1,2b] the iterative process

(22) $$A_h y^{n+1} = \lambda B_h G_h^v y^n \qquad (n \in \mathbb{N})$$

converges for any $y^o \in \mathbb{R}^{\Omega_h}$ to the unique solution y^v of

$$A_h x = \lambda B_h G_h^v x.$$

Here, the mapping G_h^v on $\mathbb{R}^{\Omega_h}$ is constructed via (17) with g replaced by g^v.

If $y^v \in [\theta, w_h]$ (which is not always the case!) then by the definition of g^v we find $g(t,y^v(t)) \leq g^v(t,y^v(t))$ $(t \in \Omega_h)$ or $G_h y^v \leq G_h^v y^v$, hence

$$\lambda\, A_h^{-1} B_h G_h y^v \leq \lambda\, A_h^{-1} B_h G_h^v y^v = y^v.$$

This proves that

(23) $$x^o = y^v, \quad A_h x^{n+1} = \lambda\, B_h G_h x^n \qquad (n \in \mathbb{N})$$

produces a sequence satisfying $\theta \leq x^{n+1} \leq x^n$ for $n \in \mathbb{N}$. Hence, x^n converges to a solution $\overline{x}_\lambda$ of (18) and we have

(24) $$\theta \le \bar{x}_\lambda \le x^{n+1} \le x^n \le w_h \qquad (n \in \mathbb{N})$$

Indeed, by Theorem 2.1 $\bar{x}_\lambda = \theta$ if $0 < \lambda < \lambda_h$. However, the next Theorem 2.2 guarantees that $\bar{x}_\lambda$ is a nontrivial solution of (18) if $\lambda_h < \lambda$.

Theorem 2.2: Let G1, G2 and G3 be satisfied. Then (22) is a globally convergent sequence with the limit y^v. If $\theta \le y^v \le w_h$ the process (23) converges to a solution $\bar{x}_\lambda$ of (18) and (24) holds. For any solution $\bar{z}_\lambda \in [\theta, w_h]$ of (18) we have $\bar{z}_\lambda \le \bar{x}_\lambda$.

In particular, $\bar{x}_\lambda$ is a nontrivial solution of (18) if and only if there exists a nontrivial solution of (18) in $[\theta, w_h]$.

Proof of Theorem 2.2: Let

$$A_h \bar{z}_\lambda = \lambda B_h G_h \bar{z}_\lambda, \qquad \theta \le \bar{z}_\lambda \le w_h$$

As above the process (22) tends to y^v if we put $y^0 = \bar{z}_\lambda$. But then

(25) $$A_h y^1 = \lambda B_h G_h^v \bar{z}_\lambda \ge \lambda B_h G_h \bar{z}_\lambda = A_h \bar{z}_\lambda$$

holds (note that $G_h^v \bar{z}_\lambda \ge G_h \bar{z}_\lambda$ by the construction of G_h^v and that $\bar{z}_\lambda \le w_h$). Now, A_h is i.m. and (25) yields $y^1 \ge \bar{z}_\lambda$. By induction we find $y^n \ge \bar{z}_\lambda$ or $y^v \ge \bar{z}_\lambda$ since y^v is the limit of y^n.

From $A_h \bar{z}_\lambda = \lambda B_h G_h \bar{z}_\lambda$, $\bar{z}_\lambda \le y^v$ we conclude by induction that the sequence (23) satisfies $\bar{z}_\lambda \le x^n \le y^v$ for all $n \in \mathbb{N}$. Since $\bar{x}_\lambda$ is the limit of x^n we finally arrive at $\bar{z}_\lambda \le \bar{x}_\lambda$. This completes the proof of Theorem 2.2.

3. Numerical results

Let us consider the discrete rotating string problem given in section 2 for h=0.1. By Theorem 2.2 we first have to calculate the limit y^v of the process (22) and to start (23) with y^v. What we rather actually do are N steps of (22) with the initial approximation $y^0 = \theta$ and we then start (23) with the last vector from (22) to perform K steps with (23) until

$$|A_h x^K - \lambda B_h G_h x^K| \leq 10^{-3}\ \delta_h.$$

In all cases we have tried the sequence x^n would converge. However, monotonicity according to (24) would not always occur as the last lines (MO) of the following tables show:

λ	2	2	3	3	3	4	4	4	4	4	4
N	1	2	1	2	3	1	2	3	4	5	6
K	16	16	7	6	6	6	5	4	4	4	4
MO	yes	yes	no	yes	yes	no	no	yes	yes	yes	yes

λ	10	10	10	10	100	100	100	100	100	100
N	1	2	3	4	1	2	3	4	5	6
K	4	3	2	2	5	4	3	2	1	1
MO	no	no	yes	yes	no	no	no	no	no	yes

We finally note the results for $\lambda=3$ and $\lambda=20$ (the fixed end of the string is $t=1$ and the free end is $t=0$).

$\lambda=3$	t	$\bar{x}_\lambda(t)$
	1.0	0.0
	0.8	0.47275
	0.6	0.83504
	0.4	1.08916
	0.2	1.23828
	0.0	1.28675

$\lambda=20$	t	$\bar{x}_\lambda(t)$
	1.0	0.0
	0.8	3.59052
	0.6	6.38229
	0.4	8.37564
	0.2	9.57105
	0.0	9.96928

Finally, we turn to the membrane problem (3a,b) with $k=0$. It is equivalent to

$$-(t^3 x')' = 2t^3(\lambda+x)^{-2}, \quad x'(0) = x(1) = 0$$

The system

$$x(0)-x(h)=0$$

$$h^{-2}(-(t-\tfrac{h}{2})^3x(t-h)+((t-\tfrac{h}{2})^3+(t+\tfrac{h}{2})^3)x(t)-(t+\tfrac{h}{2})^3x(t+h))$$

$$=2t^3(\lambda+x(t))^{-2} \quad (t=h,..,1-h)$$

$$x(1)=0$$

describes a discrete analogue on the grid Ω_h as defined in section 0. It is of the form

$$A_hx = B_hF_{\lambda h}x$$

where the matrix $A_h \in L^h$ is given by the left-hand-side of the system and where $B_h=\mathrm{diag}\,(0,1,..,1,0)\in L^h_+$. The operator $F_{\lambda h}$ on $\mathbb{R}^{\Omega_h}$ takes x onto $F_{\lambda h}x$ with the components

$$(F_{\lambda h}x)(t) = 2t^3(\lambda+x(t))^{-2} \quad (t\in\Omega_h).$$

Like the operator A in the continuous case of section 1 the matrix A_h has a nonnegative inverse. This is the key for the proofs of discrete counterparts to the Theorems 1.1 and 1.2. Since

$$\frac{d}{dv}[2t^3(\lambda+v)^{-2}] = -4t^3(\lambda+v)^{-3} \geq -4\lambda^{-3} \quad \text{for } v \geq 0$$

we have the globally convergent method

$$(26) \qquad (A_h+2\lambda^{-3}B_h)x^{n+1} = B_h(F^N_{\lambda h}x^n+2\lambda^{-3}x^n),$$

where $N\in\mathbb{N}$ and $F^N_{\lambda h}$ are constructed as in the continuous case. The next table shows the function values of the discrete solution $\bar{x}^h_\lambda$ at the grid points indicated (h=0.1) for λ=0.5. An error estimate has been done via the discrete version of the stability inequality (9). The last column of the following table shows the error of the figures in the previous column versus the true solution of the discrete system.

$\lambda=0.5$ t	$\bar{x}_\lambda^h(t)$	
0.1	0.399 705	0.0006
0.3	0.378 453	0.0001
0.5	0.327 658	0.00004
0.7	0.241 473	0.00002
0.9	0.102 617	0.000003

We remark that the process (26) is normally slow. However, it is globally convergent and produces an approximation which is a good start for Newton's method.

References

[1] Beyn, W.-J., Das Parallelenverfahren für Operatorgleichungen und seine Anwendung auf nichtlineare Randwertaufgaben, ISNM 31 (1976), 9-33.

[2a] Bohl, E., Monotonie: Lösbarkeit und Numerik bei Operatorgleichungen, Springer Tracts in Natural Philosophy, Bd. 25 (1974).

[2b] Bohl, E., Iterative procedures in the study of discrete analogues for nonlinear boundary value problems, Istituto per le applicazioni del calcolo "Mauro Picone" (1975), serie III-N. 107.

[3] Bohl, E., Lorenz, J., Inverse monotonicity and difference schemes of higher order. A summary for two-point boundary value problems, to appear.

[4a] Dickey, R. W., The plane circular elastic surface under normal pressure, Arch. Rat. Mech. Anal. 26 (1967), 219-236.

[4b] Dickey, R. W., Bifurcation problems in nonlinear elasticity, Pitman Publishing (1977).

[5] Lorenz, J., Zur Inversmonotonie diskreter Probleme, Numer. Math. 27 (1977), 227-238.

[6] Protter, M. H., Weinberger, H. F., Maximum principles in differential equations, Prentice-Hall (1967).

[7] Temme, N. M., Nonlinear Analysis, Vol. 2, Mathematisch Centrum Amsterdam (1976).

[8] Weinitschke, H. J., Verzweigungsprobleme bei kreisförmigen elastischen Platten, ISNM 38 (1977), 195-212.

A REVISED MESH REFINEMENT STRATEGY FOR NEWTON'S METHOD APPLIED TO NONLINEAR TWO-POINT BOUNDARY VALUE PROBLEMS

S. F. McCormick*

ABSTRACT

The objective of this paper is the development of a mesh refinement technique for applying Newton's method to the solution of nonlinear two-point boundary value problems. The process represents a significant improvement on an earlier approach both in its general applicability and in its numerical performance. This is demonstrated by several reported numerical examples.

*This work was supported by the National Science Foundation under grant number MCS76-09215.

I. Introduction

Consider the nonlinear two-point boundary value problem

(1) $$y'' + f(t, y, y') = 0, \ t \in [a, b]$$

with the solution subject to certain boundary conditions at t = a, b. Let $a = t_0 < t_1 < \ldots < t_{n+1} = b$ denote a sequence of partitions of [a, b] such that $h_n = \max_{0 \le i \le n} |t_{i+1} - t_i|$ tends to zero as $n \to \infty$. Suppose a discretization of (1) in the form

(2) $$F_n(\bar{y}) = \bar{0}$$

is given, where $F: \mathcal{R}^{n+2} \to \mathcal{R}^{n+2}$ acts as a discrete approximation to (1) incorporating the boundary condition so that the ith component of a solution of (2) represents an approximation to a solution of (1) evaluated at $t = t_i$.

Approximate solutions of (1) may be computed by applying Newton's method to (2), for example. With this approach, however, there are three important questions in connection with a specified discretization that must be considered, namely: What is an acceptable value for h_n so that the solutions of (1) and (2) are sufficiently close? What are the convergence criteria to be used for Newton iterations applied to (2)? Is there anything to be gained by using mesh refinement techniques to reduce the total number of operations?

The third question was treated in [1] for a special case of (1). The work was later extended in [2]. Although most of the earlier paper dealt with establishing a theoretical foundation for the phenomenon that Newton's method converges in some sense independently of the mesh size, the theory was used as a basis for a mesh refinement procedure designed to reduce the number of Newton iterates computed on the final grid. The technique proves successful in reducing overall computation although two limitations related to the first two questions noted above are evident in the strategy itself. First, a priori knowledge of the final grid must be available, something not always possible in practice. Second, the strategy is based on using a fixed convergence tolerance for each of the grid sizes used in (2). Even though convergence tolerances commensurate with the truncation error can be used, no account is made for their use in the strategy.

The object of this paper is to present an algorithm for mesh selection that overcomes these difficulties and, at the same time, significantly improves performance. Loosely speaking, the technique uses available discretization error estimates (usually attendant with unknown constants) together with some initial computation on very coarse grids to determine the size of the final mesh and the final tolerance for convergence. Backtracking from the final grid and using the coarse grid results and knowledge of the numerical behavior of Newton's method, the next-to-last grid and its convergence tolerance is determined. The object here is to "step-back" from the final grid as far as is safe to ensure the need for only one iteration on the final grid. This process is continued until a grid coarser than the one used for initialization is reached. The last iterate computed on the initial grid is then used to start the computational sequence that consists of interpolation to the finer grid followed by a single Newton iteration. This is continued until the last grid is reached and a final iteration is performed.

The notation of this paper is made somewhat difficult by the occasional need to represent vectors in $\mathcal{R}^n$ for various values of n. When this is needed in the exposition, small letters with overbar a prescript will be used. Thus, ${}_n\bar{y}$ implies ${}_n\bar{y} \in \mathcal{R}^n$, for example. The prescript will be ommitted when no ambiguity is possible.

The mesh refinement process is described in some detail in the next section. The third section illustrates the technique by discussing several numerical examples. Some concluding remarks are made in section IV.

II. The Mesh Refinement Process

The essence of the strategy developed in this section is very simple. To emphasize this, we restrict our attention to problems of the special form

$$(3) \qquad \begin{aligned} y'' + f(y) &= 0 \\ y(0) = 0 &= y(1). \end{aligned}$$

We assume that the grid consists of equally spaced points so that $h_n = \frac{1}{n+1}$. Assume also that the discretization in use is the standard

$$(4) \qquad F_n(\bar{y}) = \bar{0},$$

where $\bar{y} = (y_i) \in \mathcal{R}^n$, $F_n(\bar{y}) = \mathcal{A}_n\bar{y} + \bar{f}(\bar{y})$, $\mathcal{A}_n$ is the n by n central difference matrix approximation to the second derivative (i.e., the ith component of $\mathcal{A}_n\bar{y}$ is $(y_{i+1} - 2y_i + y_{i-1})h_n^{-2}$), and $\bar{f}(\bar{y})$ is the n-vector whose i^{th} component is $f(y_i)$. That the refinement procedure applies in a much more general setting will be apparent from the discussion that follows.

We should warn the reader that the mathematical treatment here is very loose. The inequalities and estimates rely often on ignoring higher order terms and using asymptotic estimates. Nevertheless, the results are useful in providing a foundation for the mesh refinement process. We will see in fact that the procedure is quite insensitive to the accuracy of the estimates that are made. In any case, rigor will be left to future work.

It is assumed throughout this paper that (3) exhibits a (not necessarily unique) solution $y^* \in C^2[0, 1]$ so that for some $\gamma > 0$

$$(5) \qquad \|F'(y^*)w\|_\infty \geq \gamma \|w\|_\infty$$

for all $w \in C^2[0, 1]$, where $F'(y)$ denotes the Frechet derivative of $F(y)=y''+f(y)$. Then, for sufficiently small h, by the results presented in [1] and [2], we are guaranteed that a solution $\bar{y}^*$ of (4) exists that approximates y^* arbitrarily close at the grid points. We can assume that γ is defined so that

$$(6) \qquad \|F_n(\bar{y})\|_\infty \geq \gamma \|\bar{y}\|_\infty$$

for $\bar{y}$ sufficiently near $\bar{y}^*$. We are also guaranteed that the Newton iterates given by

$$(7) \qquad \bar{y}^{(k+1)} = \bar{y}^{(k)} - F_n'^{-1}(\bar{y}^{(k)})F_n(\bar{y}^{(k)})$$

are well-defined in a neighborhood of $\bar{y}^*$ and converge quadratically for the problem in (4) with the asymptotic estimate

$$(8) \qquad \overline{\lim_{k\to\infty}} \frac{\|\bar{y}^{(k+1)} - \bar{y}^*\|_\infty}{\|\bar{y}^{(k)} - \bar{y}^*\|_\infty^2} \le \frac{1}{\gamma} \|F_n''(\bar{y}^*)\|_\infty .$$

Thus, in a small neighborhood of $\bar{y}^*$, we may again assume that γ has been defined so that

$$(9) \qquad \|\bar{y}^{(k+1)} - \bar{y}^*\|_\infty \le \frac{\|f''(y^*)\|_\infty}{\gamma} \|\bar{y}^{(k)} - \bar{y}^*\|_\infty^2 .$$

Here we have used the fact that the entries of $\bar{f}''(\bar{y}^*)$ are approximations of $f''(y^*)$ at the grid points.

The inequality in (9) provides information concerning Newton's method that will be useful in developing the mesh refinement procedure. Equally important is a knowledge of the behavior of the discretization defined in (4). To this end, let $\bar{y} = P_n y^*$, where P_n: $C^2[a, b] \to \mathcal{R}^n$ is defined so that the ith component of $P_n y$ is equal to $y(t_i)$ with $t_i = \frac{i}{n+1}$. Then simple Taylor series estimates yield

$$(10) \qquad F_n(\bar{y}) = \bar{\delta}_n,$$

where $\|\bar{\delta}_n\|_\infty \le \frac{1}{12}\|y^{*IV}\|_\infty h_n^2$. Hence,

$$\|\bar{y} - \bar{y}^*\|_\infty = \|F_n'^{-1}(\bar{y}^*)F_n'(\bar{y}^*)(\bar{y} - \bar{y}^*)\|_\infty$$

$$= \frac{1}{\gamma}\|F(\bar{y}^*) + F_n'(\bar{y}^*)(\bar{y} - \bar{y}^*)\|_\infty .$$

Note that $F_n(\bar{y}^*) + F'(\bar{y}^*)(\bar{y} - \bar{y}^*) \cong F_n(\bar{y})$. Hence, in a small neighborhood of $\bar{y}^*$, we may as well assume γ is defined so that

$$(11) \qquad \|\bar{y} - \bar{y}^*\|_\infty \le \frac{\|y^{*IV}\|_\infty}{12\gamma} h_n^2 .$$

Inequalities (9) and (11) are all that is needed to set up the mesh refinement process. For suppose an $\varepsilon > 0$ is given with the requirement that a vector $\bar{y} \in \mathcal{R}^n$ be computed so that the piecewise linear function interpolating $\bar{y}$ at the grid points is within ε of a solution of (1). Then from (11) we can determine how fine the grid must be to guarantee that our discrete answer is within, say, $\varepsilon/2$ of some y^*. The objective is then to produce a vector in $\mathcal{R}^n$ within $\varepsilon/2$ of $\bar{y}^*$. By (9) we are able to determine how close the initial guess for Newton's method must be on this final grid to get within $\varepsilon/2$ of $\bar{y}^*$ in one iteration. This then gives us an estimate for the required truncation error on the previous grid. Stepping back in this way is the essence of the mesh refinement process.

Of course, before this can be done in practice, there are quantities in (9) and (11) that must be approximated in some sense. The quantities $||f''(y^*)||_\infty$ and $||y^{*IV}||_\infty$ may be approximated either by analytical means, by obtaining an approximation to y* on a fairly coarse grid and estimating these quantites, or by a combination of these two approaches. For the numerical experiments cited in the next section, sufficient analytical information is available to provide excellent estimates for our purposes. In any event, the approximation of these quantities is not a critical problem.

Approximation of γ can be made by solving (4) on a coarse grid, observing the successive Newton iterates, and choosing a sufficiently conservative upper bound for the computed ratios

(12) $$\frac{||\bar{y}^{(k+1)} - \bar{y}^{(k)}||_\infty}{||\bar{y}^{(k)} - \bar{y}^{(k-1)}||_\infty^2},$$

where $k > 0$ is taken over the allowable range of the iterations actually performed. (It is probably best to stop the iterations on the initial grid when the change in the approximate vector is near machine precision.) Then, using the fact that the difference of successive iterates of a quadratically convergent method is nearly equal to the actual error, this upper bound together with (9) and an estimate for $||f''(y^*)||_\infty$ provide the necessary approximation to γ.

We can assume, then, that some work has already been expended on an initial coarse grid to determine quantities q_1 and q_2 that satisfy

(13) $$||\bar{y}^{(k+1)} - \bar{y}^*||_\infty \leq q_1 \, ||\bar{y}^{(k)} - \bar{y}^*||_\infty^2$$

and

(14) $$||\bar{y} - \bar{y}^*||_\infty \leq q_2 h_n^2.$$

We let n_0 denote the number of initial grid points interior to [0, 1] and $\bar{w} = {}_{n_0}\bar{w}$ the final Newton iterate computed on the initial grid.

Assume that $p \geq 1$ grids are to be used in the mesh refinement process. (Of course, p is yet to be determined. It is needed here for notational purposes, however.) From (14) and with $\varepsilon > 0$ given, the final grid is now determined according to the requirement

(15) $$q_2 h_{n_p}^2 < \varepsilon/2.$$

That is, the smallest number of grid points, n_p, that should be used for the final grid is given by

(16) $$n_p = [(\frac{2q_2}{\varepsilon})^{1/2} - 1].$$

Here, [x] denotes the least integer greater than or equal to x. Moreover, to ensure that the convergence criteria for the final grid is commensurate with the truncation error, we use

(17) $$\varepsilon_p = q_2 h_{n_p}^2.$$

Note that

(18) $$\varepsilon_p < \varepsilon/2.$$

Using (16) guarantees that the discrete solution is within $\varepsilon/2$ of the true solution of (1) and with (18) we know that the final computed approximation is within $\varepsilon/2$ of the discrete solution. This ensures as intended that the computed approximation is within ε of the true solution of (1).

In order that only one iteration is required on the final grid, by (13) it suffices to have an initial approximation $\bar{y}^{(0)} = {}_{n_p}\bar{y}^{(0)}$ that satisfies

(19) $$q_1 \|\bar{y}^{(0)} - \bar{y}^*\|_\infty^2 < \varepsilon_p.$$

Ignoring interpolation effects, then (19) is satisfied if the Newton iterate on grid p - 1 is within $q_2 h_{n_{p-1}}^2$ of ${}_{n_{p-1}}\bar{y}^*$ and if

(20) $$q_1 (2q_2 h_{n_{p-1}}^2)^2 < \varepsilon_p.$$

Here we used the fact that $\|\bar{y}^{(0)} - \bar{y}^*\|_\infty \cong \|\bar{y}^{(0)} - P_{n_p} y^*\|_\infty$. Since $h_{n_{p-1}} = \frac{1}{n_{p-1}+1}$, it suffices that

(21) $$n_{p-1} = [(\frac{4q_1 q_2^2}{\varepsilon_p})^{1/4} - 1].$$

Here again we use the convergence tolerance

(22) $$\varepsilon_{p-1} = q_2 h_{n_{p-1}}^2.$$

Note by (17) that (21) can be rewritten as

$$n_{p-1} + 1 = [K(n_p + 1)^{1/2}],$$

where $K = (4q_1 q_2)^{1/4}$. This says in essence that, up to the factor K, the best refinement process is achieved by squaring the present mesh size. This is, of course, directly attributable to the quadratic convergence property of Newton's method. The truncation error for the discretization process has its effects in determining the value of K, only.

The full mesh refinement process is now easily derived. With n_p determined by (16) and n_{p-1} by (23), coarser meshes are given according to

(24) $$n_{i-1} + 1 = [K(n_i + 1)^{1/2}], \quad i = p, p-1, \ldots, 2, 1.$$

Here, p is determined so that $K(n_1+1)^{1/2} < n_0+1 \leq n_1+1$, where n_0 is the number of points on the initial grid used to determine q_1, q_2 and $\bar{w}$. In each case, the convergence tolerance is

(25) $$\varepsilon_i = q_2 h_{n_i}^2 \quad (= \frac{q_2}{(n_i + 1)^2}), \qquad i = p, p-1, \ldots, 2, 1.$$

The process is computed by interpolating $\bar{w}$ to form $_{n_1}\bar{y}^{(0)}$, performing one Newton iteration to determine $_{n_1}\bar{y}^{(1)}$ in (7), interpolating $_{n_1}\bar{y}^{(1)}$ to form $_{n_2}\bar{y}^{(0)}$, and continuing in this way until the final result, $_{n_p}\bar{y}^{(1)}$, has been computed.

To summarize, the following steps of the complete mesh refinement process are listed:

STEP I: Choose an initial uniform grid with a relatively small number of points, n_0, interior to [0, 1]. Perform the iterations in (7) until the relative error

$$\frac{\|\bar{y}^{(k+1)} - \bar{y}^{(k)}\|_\infty}{\|\bar{y}^{(k)}\|_\infty}$$

is near machine accuracy. Determine a conservative upper bound for the ratios computed in (12), denoting this bound by γ. Approximate the quantities $\|f''(y^*)\|_\infty$ and $\|y^{*IV}\|_\infty$ in (9) and (11), respectively, by analytical means if possible. Otherwise, use $\bar{y}^{(k+1)}$ as an approximation to y^* and note that

$$\|f''(y^*)\|_\infty \cong \|\bar{f}''(\bar{y}^{(k+1)})\|_\infty$$

and

$$\|y^{*IV}\|_\infty \cong \|\mathcal{A}_{n_0} \bar{f}(\bar{y}^{(k+1)})\|_\infty .$$

Now determine the quantities in (13) and (14) according to

$$q_1 = \frac{\|f''(y^*)\|_\infty}{\gamma}$$

and

$$q_2 = \frac{\|y^{*IV}\|_\infty}{12\gamma} .$$

STEP II: Let n_p be given by (16) and define $n_{p-1}, n_{p-2}, \ldots, n_0$ according to (24). The integer $p \geq 1$ is defined so that $K(n_1 + 1)^{1/2} \leq n_0 + 1 < n_1 + 1$, that is, so that the results on the initial grid of n_0 points are adequate to start the process.

STEP III: Let $i = 0$.

STEP IV: Interpolate $_{n_i}\bar{y}^{(1)}$ linearly to form a vector in $\mathcal{R}^{n_{i+1}}$ which we denote by $_{n_{i+1}}\bar{y}^{(0)}$. Perform one Newton iteration to form $_{n_{i+1}}\bar{y}^{(1)}$.

STEP V: Increment i by one. If $i = p$, accept $_{n_p}\bar{y}^{(1)}$ as the final approximation and stop. Otherwise, repeat step 4.

III. Numerical Results

In this section we report on experiments with the strategy described in section II applied to the three problems discussed in [1], namely:

P1
$$y'' + y^3 = 0$$
$$y(0) = 0 = y(1)$$
$$y(t) \geq 0.$$

P2
$$y'' + e^y = 0$$
$$y(0) = 0 = y(1)$$
upper solution

P3
$$y'' + 2\pi^2 \sin y = 0$$
$$y(0) = 0 = y(1)$$
$$y(t) \geq 0$$

The ancillary conditions in each case above indicate the solution we attempted to compute from among the multiple solutions of each problem. Note by implication that P2 exhibits two solutions and that one (upper) is strictly greater than the other in (0, 1).

In each case, experimentation began with $n_0 = 2^5$. For comparison purposes, ε was chosen so that ε_p would always be 2^{12}. Note that the determination of $\|f''(y^*)\|_\infty$ and $\|y^{*IV}\|_\infty$ can be done analytically using the facts that $y^{*''}=-f(y^*)$, $\|y^*\|_\infty = y(1/2)$, and that $y'(t) = 0 \Rightarrow t = 0$. Available estimates for $y(1/2)$ such as in [3] then provide accurate estimates for these quantites.

Quantites determined in the process are listed in table 1. Note that $p = 3$ in each case. Operation counts for each together with those observed using the strategy in [1] are given in table 2.

$\|f''(y)\|_\infty$	$\|y^{IV}\|_\infty$	K	γ	q_2	q_1	h_1	h_2	h_3	ε_1	ε_2	ε_3	ε	
22	2080	1.68	44	.5	4	2^{-5}	2^{-7}	2^{-12}	4E-3	2.5E-4	2.5E-7	5E-7	P1
60	3641	1.83	80	.75	3.75	2^{-5}	2^{-7}	2^{-12}	3.7E-3	2.3E-4	2.3E-7	4.6E-7	P2
20	186	.95	40	.5	.4	2^{-5}	2^{-7}	2^{-12}	4E-4	E-4	2.5E-8	5E-8	P3

Table 1. Computed Parameters for P1, P2, P3

MULTIPLY/DIVIDES

Problem	Present Strategy	Previous [1] Strategy
P1	21,555	31,974
P2	21,555	31,974
P3	21,235	31,638

Table 2. Observed operation counts (MULTIPLY/DIVIDES) for P1, P2, P3 for the present and previous strategies

IV. Concluding Remarks

In the paper we have developed a complete mesh refinement process using Newton's method for solving nonlinear two-point boundary value problems. Although the development was not rigorous, the technique has shown great success as demonstrated by the numerical results reported in section III. Moreover, although the setting was restricted to a special class of problems, it is clear that the process has general applicability. Truncation error estimates for the discretization procedure are the major requirements in this regard.

There are two pitfalls that must be considered with this approach, however. First, there is no guarantee that the computed γ actually satisfies the condition introduced in (5). To be safe, an extra iteration on the final grid may sometimes be required to assure that convergence has occurred. Second, the convergence of Newton's method itself requires acceptable starting guesses. Poor initialization may corrupt the efforts made in this process. However, this is just the very purpose of the mesh refinement technique. In fact, the only real concern is that the grid chosen for the initial computation is both coarse enough to allow for possible intensified computational efforts, yet fine enough to exhibit a solution that loosely approximates the true solution of (1). (In this last regard, for example, multiple solutions of (1) will not be separated in (4) if n_0 is chosen too small.) With n_0 suitably chosen, we are then able to exert as much effort as is necessary to single out the target solution on the initial grid. The process can be viewed in this way as an alternative to the usual methods for overcoming the initial guess difficulties inherent in Newton's method.

REFERENCES

[1] E. L. Allgower and S. F. McCormick, A phenomenon concerning Newton's method for boundary value problems and its application to mesh refinement, Numer. Math., to appear.

[2] E. L. Allgower, S. F. McCormick, and D. V. Pryor, A general mesh independence principle for Newton's method applied to second order boundary value problems, submitted for publication.

[3] E. L. Allgower, On a discretization of $y'' + \lambda y^k = 0$, Topics in Numerical Analysis II, ed. J. J. H. Miller, New York, Academic Press, (1975) 1-15.

Problems in applying the SOR-method to the solution of the Maxwell's time dependent equations

W.ELBEN and W.WOLFF

1. Introduction

The mathematical problem is that of solving Maxwell's equations for a known time dependence and a given distribution of current and material. The material can be electrically conductive and ferromagnetic.

Ignoring the displacement current density Maxwell's equations are:

(1) $\quad \mathrm{curl}\ \underline{H} = \underline{S}$

(2) $\quad \mathrm{curl}\ \underline{E} = -\dfrac{d}{dt}\underline{B}$

(3) $\quad \mathrm{div}\ \underline{B} = 0$

with

$\underline{S}$ = current density

$\underline{B}$ = magnetic flux density

$\underline{H}$ = magnetic field strength

$\underline{E}$ = electric field strength .

The current density consists of a given part $\underline{S}_L$ and an induced part $\underline{S}_w$

$$\underline{S} = \underline{S}_L + \underline{S}_w$$

with $\quad \underline{S} = \sigma\underline{E}$

σ = electric conductivity .

The electrical conductivity can be represented by a diagonal matrix independent of the electric field.

The relationship between magnetic flux density and the magnetic field strength is given by

$$\underline{B} = \mu(H)\ \underline{H} .$$

The permeability is a diagonal matrix with elements depending on the magnetic field strength. Where time dependent fields are concerned, the permeability is assumed to depend on the maximum field strength. Regarding these assumptions, Maxwell's equations can be written in the following form:

$$(4) \qquad \text{curl } \underline{H} = \underline{S}_L + \underline{S}_w$$

$$(5) \qquad \text{curl } \sigma^{-1} \underline{S}_w = -\mu\,(\hat{H})\,\frac{d}{dt}\,\underline{H}$$

$$(6) \qquad \text{div } \mu \underline{H} = 0 \; .$$

$\underline{S}_L$, σ and μ are given quantities of locus. Unknown are the vektorfields $\underline{H}$ and $\underline{S}_w$ fulfilling the equations (4) – (6).

The permeability is always greater than zero. The electrical conductivity is zero outside the electrically conductive material, therefore equation (5) is valid for the interior space only. Because the eddy current density $\underline{S}_w$ is zero outside the electrical conductors, equation (5) must be solved in the interior space only. Assuming the conductors with homogenous σ are surrounded by non conducting material only, equation (5) can be multiplied by σ :

$$(7) \qquad \text{curl } \underline{S}_w = -\sigma\mu\frac{d}{dt}\underline{H} \; .$$

Further assuming that the given current and the induced current are not mixed at all the following equation

$$(8) \qquad \text{curl } \underline{H} = \underline{S}_w$$

is valid in the interior space.

Eliminating $\underline{S}_w$ in equation (7) and (8) the following relationship is valid for the magnetic field strength in the interior space

$$(9) \qquad \text{curl curl } \underline{H} + \sigma\mu\frac{d}{dt}\,\underline{H} = 0 \; .$$

In the exterior space the well known equations of the stationary case are valid:

$$(10) \qquad \text{curl } \underline{H} = \underline{S}_L$$

$$(11) \qquad \text{div } \mu\underline{H} = 0 \; .$$

For solving equations (10) and (11), Sommerfeld /1/ split the magnetic field strength into two parts

$$(12) \qquad \underline{H} = \underline{H}_i + \operatorname{grad} \Phi$$

with

$$(13) \qquad \operatorname{curl} \underline{H}_i = \underline{S}_L$$

and

$$(14) \qquad \operatorname{div} \mu \operatorname{grad} \Phi = -\operatorname{div} \mu \underline{H}_i \;.$$

The application of equation (12) to the computation of stationary magnetic fields has already described in detail (/2/, /3/, /4/). Because the interior space is surrounded by nonconductors the following condition is valid at the interfaces

$$(15) \qquad \operatorname{curl}_n \underline{H} = \underline{S}_{w,n} = 0 \;.$$

This condition is fulfilled if the tangential component of $\underline{H}$ can be expressed by a gradient of a scalar potential in the interfaces. That means : the boundary belongs to the exterior space.

In order to describe the problem two system quantities are introduced : in the exterior space including the boundary, the scalar potential; and in the interior space, the magnetic field strength. At the boundary both systemquantities are connected by the divergence condition.

2. Certain time dependencies

The time dependence is known for many practical problems. In this case the time can be eleminated and a problem depending on spatial coordinates only has to be solved.

In most practical problems the time dependence is sinusoidal

$$(16) \qquad \mu \underline{H} = \sum_{\nu=-N}^{N} \tilde{\mu}_\nu \underline{\tilde{h}}_\nu \, e^{j\omega\nu t}$$

$\tilde{\mu}_\nu$, $\underline{\tilde{h}}_\nu$ are the complex Fourier coefficients.

Provided all the $\tilde{\mu}_\nu$ equal $\bar{\mu}$, and $\bar{\mu}$ depends on locus only, the time dependence of the magnetic field strength is the same as the timedependence of the magnetic flux density. For each Fourier coefficient the destination equation is valid

(17) $\quad \text{curl curl } \underline{\tilde{h}}_\nu + j\sigma\omega\tilde{\mu}\underline{\tilde{h}}_\nu = 0$

with

$$\underline{H} = \sum_{\nu=-N}^{N} \underline{\tilde{h}}_\nu \, e^{j\nu\omega t} \quad .$$

In most devices with alternating flux density, magnetic material with negligible hysteresis is used. In this case by dividing the maximum values of the flux density and the magnetic field strength one gets $\tilde{\mu}$.

3. Derivation of the difference equations

In order to derive the difference equations a grid is placed over the total computation space.

$x = x_i$, $y = y_j$ and $z = z_k$

$i = 1,\dots,Nx$, $j = 1,\dots,Ny$, $k = 1,\dots,Nz$.

At the mesh points of the exterior space and at the boundary of the interior space the values of the scalar potential Φ are calculated by numerical solution of equation (14). The components of the total magnetic field are computed at the centerpoints between two mesh points by means of equation (12). The derivatives $\frac{\partial\Phi}{\partial x}$, $\frac{\partial\Phi}{\partial y}$ and $\frac{\partial\Phi}{\partial z}$ are approximated by the difference of Φ-values at the neighbouring points divided by the distance.

In the same way one can choose the calculation points for the components of $\underline{H}$ in the interior space, i.e. at the centerpoints between two mesh points. Note that the components of the magnetic field $\underline{H}$ are calculated at different points for the x-, y- and z-directions.

The mesh points are numbered continously. The calculation points of the components H_x, H_y, H_z of $\underline{H}$ equal the number at the left, front and lower neighbouring mesh point (fig. 1), respectively.

The z-component of equation (17) can be written in the following form

$$(18) \qquad \frac{\partial^2 H_z}{\partial x^2} + \frac{\partial^2 H_z}{\partial y^2} - \frac{\partial}{\partial z}\left(\frac{\partial H_x}{\partial x} + \frac{\partial H_y}{\partial y}\right) - j\sigma\omega\mu H_z = 0 \quad .$$

The discretization of the two dimensional Laplace term $\frac{\partial^2 H_z}{\partial x^2} + \frac{\partial^2 H_z}{\partial y^2} = 0$ results in an expression of the form /5/

$$\sum_{v=1}^{4} a_v H_{z,v} - H_{z,o} \cdot \sum_{v=1}^{4} a_v = 0$$

where $H_{z,v}$ is the value of H_z at the point P_v (fig. 2).
If the term $-j\sigma\omega\mu H_{z,o}$ of equation (18) is added, where $H_{z,o}$ is the value of H_z at the central point P_o, the following is obtained

$$\sum_{v=1}^{4} a_v H_{z,v} - H_{z,o} \left(\sum_{v=1}^{4} a_v + j\sigma\mu\omega\right) = 0 .$$

This term increases the diagonal dominance and therefore improves the convergence of the iterative solution.
For the remaining term of (18) one obtains expressions of the form

$$\sum_{v=1}^{4} b_v H_{x,v} \quad \text{and} \quad \sum_{v=1}^{4} c_v H_{y,v}$$

which produce off-diagonal elements only. In extreme casis these terms disturb the diagonal dominance.
For the points at the boundary of the interior space special considerations are necessary. The derivation of the difference equations for the components H_x, H_y, H_z at these points is shown by means of an example (fig. 2).
Assuming that the component H_z is calculated at point P_o and that the point P_4 lies at the boundary of the interior space, then $H_{z,4}$ is calculated by equation (12) as

$$H_{z,4} = H_{iz,4} + \left(\frac{\partial}{\partial z}\Phi\right)_{P_4} \simeq H_{iz,4} + g\,(\Phi_{v,1} - \Phi_{v,2}) .$$

Thus the difference equations for the boundary points contain in addition the potential Φ. Therefore the values of Φ calculated in the exterior space are included in the linear equation system for the components H_x, H_y, H_z of $\underline{H}$.
The discretization of the divergence condition at the interfaces is shown again by means of an example (fig. 3). The application of the five-point formula in three dimensions to the potential equation (14) gives an expression of the form

$$(19) \qquad \sum_{v=1}^{6} h_v \left(H_{i,v} + \frac{\Phi_v - \Phi_o}{d_v}\right) = 0 .$$

For $v = 1,2,4,5,6$ the expression

$$H_{i,v} + \frac{\Phi_v - \Phi_o}{d_v}$$

can be calculated. If $v = 3$ (fig. 3) this expression has to be replaced by $-H_{z,3}$ because $\overline{P}_3$ is the calculation point of H_z in the interior space.
Thus one gets expressions of the form

$$\sum_{\substack{v=1\\v\neq 3}}^{6} h_v \Phi_v - \Phi_o \sum_{\substack{v=1\\v\neq 3}}^{6} h_v + h_3 H_{z,3} = K \ .$$

By combining the above a linear equation system of the unknowns

$$\Phi_1 \ldots \Phi_N,\ H_{x,1} \ldots H_{x,Nx},\ H_{y1} \ldots H_{y,Ny},\ H_{z,1} \ldots H_{z,Nz}$$

results.
The associated coefficient matrix can be represented as shown in fig. 4.

Outside the marked areas the elements of the matrix are zero. Three types of submatrices with following properties may be distinguished:

type A_1 : real, symmetric, diagonal dominant,irreducible, consistently ordered, property A.

type A_2 : complex diagonal elements, usually not diagonally dominant.

type A_3 : complex diagonal elements, usually diagonally dominant.

4. Iterative solution of the linear system

Where magnetostatic problems are concerned, the matrix of the difference equations has type A_1. This kind of linear systems can be successfully solved by means of the successive overrelaxation (SOR) method. We also apply the SOR method to linear systems with matrices as shown in Fig. 4. In this case we use different relaxation factors for the submatrices A_1, A_2 and A_3, where the optimum relaxation factor ω_{best} for the matrix type A_1 is determined by Young. We are not so much concerned with isolated cases, where the SOR methoddoes not converge at all. Instead, our problem lies in the large number of iterations and therefore in the large computation time required for a given error reduction, as there is no method to determine the optimum relaxation factor for the matrices of type A_2 or type A_3. We believe, by determining these optimum relaxations factors a significant improvement of the computation time would be achieved.

We shall study in this final chapter the problems arising in applying the SOR method to linear systems with matrices of type A_3 which are diagonally dominant and possess complex diagonal elements.

If the matrix A belongs to a certain class of matrices, then there exists a relation between the eigenvalues of the matrix L_ω associated with the SOR method and the eigenvalues of the matrix B associated with the Jacobi method. This is the well known Young /6/, /7/ theorem :

Let A be a consistently ordered matrix with nonvanishing diagonal elements. If $\omega \neq 0$, real, and if λ is an nonzero eigenvalue of L_ω, and if μ satisfies the relation

(20) $$(\lambda + \omega - 1)^2 = \omega^2 \mu^2 \lambda$$

then μ is an eigenvalue of B. On the other hand if μ is an eigenvalue of B, and if λ satisfies (20), then λ is an eigenvalue of L_ω. Equivalent to equation (20) is the following equation /6/, /7/:

(21) $$\lambda + \omega - 1 = \omega \mu \lambda^{1/2} .$$

The relation (21) is a mapping between the complex μ- and $\lambda^{1/2}$-planes. This mapping has the following properties /7/:

If $\omega \neq 0$ and if $\rho^2 \neq |\omega - 1|$, then the circle $|\lambda^{1/2}| = \rho$ in the $\lambda^{1/2}$-plane is mapped on the ellipse

(22) $$E_{\rho,\omega} : \frac{\mu_1^2}{\frac{1}{\omega^2}(\rho + \frac{\omega-1}{\rho})^2} + \frac{\mu_2^2}{\frac{1}{\omega^2}(\rho - \frac{\omega-1}{\rho})^2} = 1$$

in the μ-plane, where $\mu = \mu_1 + i\mu_2$.

Conversely, if $\rho^2 \neq |\omega - 1|$ the ellipse $E_{\rho,\omega}$ in the μ-plane is mapped on the two circles

$$|\lambda^{1/2}| = \rho$$

and $$|\lambda^{1/2}| = \frac{\omega - 1}{\rho}$$

in the $\lambda^{1/2}$-plane.

If $\rho^2 = |\omega - 1|$, then the two circles $|\lambda^{1/2}| = \rho$ and $|\lambda^{1/2}| = \frac{|\omega - 1|}{\rho}$ coincide and are mapped on the segment $|\mu_1| \leq \frac{2\sqrt{\omega - 1}}{\omega}$, $\mu_2 = 0$, if $\omega \geq 1$ and on the segment $\mu_1 = 0$, $|\mu_2| \leq \frac{2\sqrt{1-\omega}}{\omega}$, if $\omega < 1$.

If we choose $\rho = 1$ in relation (22), then the ellipse

$$E_{1,\omega} : \mu_1^2 + \frac{\mu_2^2}{(\frac{2-\omega}{\omega})^2} = 1$$

in the μ-plane is mapped on the two circles

$$|\lambda^{1/2}| = 1$$

$$|\lambda^{1/2}| = |\omega - 1|$$

in the $\lambda^{1/2}$-plane.

With these properties of the mapping (21) Young /7/ has proved the following theorems:

Theorem 1 : Let A be a consistently ordered matrix with nonvanishing diagonal elements.

If for some positive number D all eigenvalues $\mu = \mu_1 + i\mu_2$ of the Jacobi-matrix B belong to the interior of the ellipse

$$\mu_1^2 + \frac{\mu_2^2}{D} = 1 ,$$

then the SOR method convergenes for any ω in the range

$0 < \omega < \frac{2}{1+D}$.

Theorem 2 : Let A be a consistently ordered matrix with nonvanishing diagonal elements.

If the SOR method converges, then there exists a positive number D such that all eigenvalues $\mu = \mu_1 + i\mu_2$ of the Jacobi-matrix B lie inside the ellipse

$$\mu_1^2 + \frac{\mu_2^2}{D^2} = 1 .$$

Well known is a special case of these two theorems:

Theorem 3 : If A is a consistently ordered matrix with nonvanishing diagonal elements such that the Jacobi-matrix B has real eigenvalues, then the SOR method converges if and only if

$0 < \omega < 2$ and $\rho(B) < 1$, where $\rho(B)$

is the spectral radius of B.

It follows from theorem 3 that the SOR method applied to the system with matrix type A_1 converges.

For this type of linear systems the value of ω_{best} which is optimum in the sense of minimizing the spectral radius $\rho(L_\omega)$ is determined by Young /6/, /7/:

Theorem 4 : Let A be a consistently ordered matrix with nonvanishing diagonal elements such the Jacobi-matrix B has real eigenvalues and such that $\rho(B) < 1$. Then

$$\omega_{best} = \frac{2}{1+\sqrt{1-\rho(B)^2}}$$

and

$$\rho(L_\omega) > (L_{\omega_{best}}) \qquad \text{for } \omega \neq \omega_{best}$$

and

$$\rho(L_\omega) = \omega - 1 \qquad \text{for } \omega_{best} \leq \omega < 2 \ .$$

Therefore if the matrix of the linear system is of the type A_1 the SOR method is optimal applicable, i.e. the SOR method converges and ω_{best} is determined.

Let us now consider the application of the SOR method to linear systems with matrices of type A_3. Assuming the known distribution of the eigenvalues μ for the Jacobi-matrix B, theorem 1 determines the convergence of the SOR method. However, in practice little is known about the distribution of eigenvalues μ. There are three problems to be solved:

P1 Is it possible for a matrix of type A_3 to determine ellipses in the μ-plane, which contain the eigenvalues μ of the Jacobi-matrix B ?

P2 Given an ellipse in the μ-plane, containing the eigenvalues of the B-matrix, what is the value of ω which minimizes the radius of the circle in the $\lambda^{1/2}$-plane on which the ellipse is mapped ?
What is the value of this radius ?

P3 Taken that the μ are contained within certain ellipses in the μ-plane and taken that we know the optimum value of ω for each of these ellipses, which is the "best" ellipse, i.e. which one minimizes the radius of the circle in the $\lambda^{1/2}$-plane ?

In a special case the problem P2 is solved by Young /7/:

Theorem 5 : Let A be a consistently ordered matrix with nonvanishing diagonal elements. If for all eigenvalues μ of B $|Re\,\mu|<1$ and if

$$v_1 = \max_{\mu}(Re\,\mu) \quad \text{and} \quad v_2 = \max_{\mu}(IM\,\mu)$$

are eigenvalues of B and if all eigenvalues $\mu = \mu_1 + i\mu_2$ of B lie in the closed interior of the ellipse

$$\frac{\mu_1^2}{v_1^2} + \frac{\mu_2^2}{v_2^2} = 1$$

then

$$\omega_{best} = \frac{2}{1+\sqrt{1-(v_1^2-v_2^2)}}$$

and

$$\rho(L_{best}) = \frac{v_1+v_2}{|v_1-v_2|}\,|\omega_{best}-1| = \frac{(v_1+v_2)^2}{(1+\sqrt{1+v_1^2-v_2^2})^2} .$$

For purposes of convenience we define an "eigenvalue-ellipse" of B as an ellipse centered on the origin of the μ-plane, which contains in its closed interior all eigenvalues of B and at least one eigenvalue lying on the ellipse.

According to theorem 5, Young has determined ω_{best} and $\rho(L_{\omega_{best}})$ for all eigenvalue ellipses E(a,b) of B where $\mu = a$ and $\mu' = ib$ are eigenvalues of B . The following question arises from the above mentioned cases: What happens, when the assumptions of theorem 5 are not fulfilled but when an eigenvalue-ellipse $E(|a|,|b|)$ exists centred on the origin of the μ-plane, whose major semiaxis $|a|$ form an angle β with the μ_1-axis ($\mu = (\mu_1,\mu_2)$), and when a and b of $E(|a|,|b|)$ are eigenvalues of B ? What can be said about the convergence conditions for the SOR method and the ω_{best} and $\rho(L_{\omega_{best}})$?

It can easily be seen that an exact copy of Young's theorem with the relations (20) and (21) is valid in the case of complex ω . The proof is the same as in /6/, /7/.

If ω complex, $\omega \neq 0$, $\omega \neq 1$ and if $\rho^2 \neq |\omega - 1|$ then the ellipse

$$(23) \qquad E_{\rho,\omega} : \frac{(\mu_1 \cos\beta - \mu_2 \sin\beta)^2}{\frac{1}{|\omega|^2}(\rho + \frac{|\omega-1|}{\rho})^2} + \frac{(\mu_1 \sin\beta + \mu_2 \cos\beta)^2}{\frac{1}{|\omega|^2}(\rho - \frac{|\omega-1|}{\rho})^2} = 1$$

in the μ-plane is mapped by the relation (21) on the two circles

$$|\lambda^{1/2}| = \rho$$

and $$|\lambda^{1/2}| = \frac{|\omega - 1|}{\rho}$$

in the $\lambda^{1/2}$-plane, where $\beta = \arg\left(\frac{\omega}{\sqrt{\omega - 1}}\right)$.

(23) is the equation of an ellipse with center at the origin of the μ-plane which has the semiaxes $\frac{1}{|\omega|^2}(\rho \pm \frac{|\omega-1|}{\rho})^2$ and the larger semiaxes of which forms an angle β with the μ_1-axis in the μ-plane.

If we use $\rho = 1$ in relation (23), then the following theorem is valid /8/:

Theorem 6 : Let A be a consistently ordered matrix with nonvanishing diagonal elements.
If $\rho = 1$ and if all eigenvalues of the Jacobi-matrix B belong to the interior of the ellipse $E_{1,\omega}$ as defined in (23) and if $|\omega - 1| < 1$, then the complex successive overrelaxation, i.e. SOR method with ω complex, converges.

Theorem 6 determines the convergence of the complex SOR method. If we permit complex ω, then we can extend theorem 4 as follows /8/:

Theorem 7 : Let A be a consistently ordered matrix with nonvanishing diagonal elements.
If all eigenvalues of the Jacobi-matrix B belong to a straight-line through the origin of the μ-plane and if $\tilde{\mu}$ is the eigenvalue of B with largest absolute value, i.e. spectral radius $\rho(B) = |\tilde{\mu}|$ and if $\rho(B) < 1$ then

$$\omega_{best} = \frac{2}{1 + \sqrt{1 - \tilde{\mu}^2}}$$

and

$$\rho(L_\omega) > \rho(L_{\omega_{best}}) \text{ if } \omega \neq \omega_{best}$$

and

$$\rho(L_{\omega_{best}}) = |\omega_{best} - 1| .$$

Theorem 7 solves the problems P2 and P3 only when the eigenvalue-ellipse is a straight-line through the origin of the μ-plane.

If $E(|a|,|b|)$ is an eigenvalue-ellipse of B then it results from (23) that a and b determine ω_o and $\rho(L_{\omega_o})$ as follows:

Theorem 8 : Let A be a consistently ordered matrix with nonvanishing diagonal elements.

If $E(|a|,|b|)$, $|a| \neq |b|$, is an eigenvalue-ellipse of B and if

$$\omega_o = \frac{2}{1+\sqrt{1-(a^2+b^2)}}$$

then

$$\rho(L_{\omega_o}) = \frac{|a|+|b|}{\big||a|-|b|\big|} \, |\omega_o - 1| \, .$$

We can see that if $b = 0$ then $\omega_o = \omega_{best}$ (theorem 4 and theorem 7). Furthermore if $a = |a|$ and $b = i|b|$ are eigenvalues, then $\omega_o = \omega_{best}$ (theorem 5). We believe, that this is true for all the eigenvalue-ellipses of B , i.e. $\omega_o = \omega_{best}$. If we could proof this, problem P2 would no longer exist.

We could determine ω_{best} and $\rho(L_{\omega_{best}})$ for every eigenvalue-ellipse of B .

Problems P1 and P2 are yet to be solved:

Is it possible to determine eigenvalue-ellipses of B for a matrix of type A_3, and which of them is "best" as far as minimizing $\rho(L_\omega)$ is concerned ?

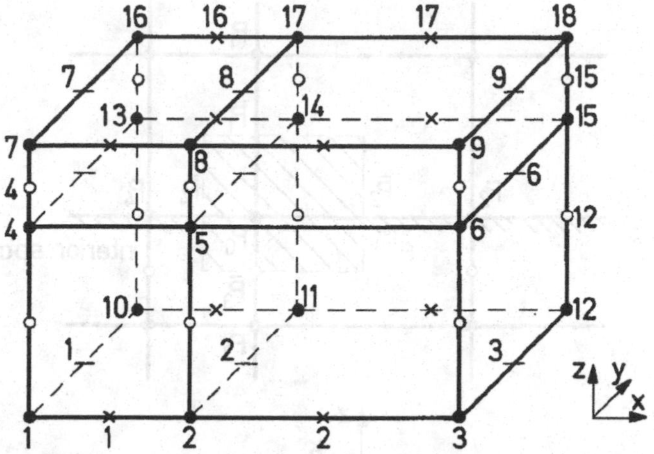

Fig.:1 Ordering of the mesh points and calculation points

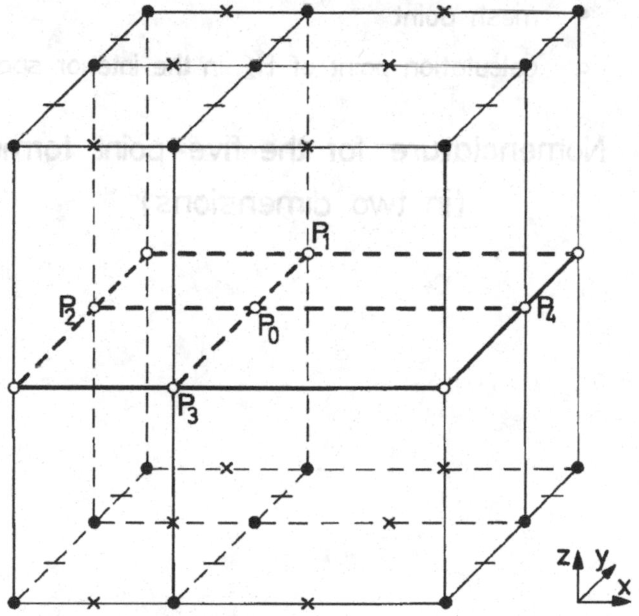

- ● mesh point
- × calculation point of H_x
- – calculation point of H_y
- ○ calculation point of H_z

Fig.:2 Nomenclature at a calculation point P_0 of H_z

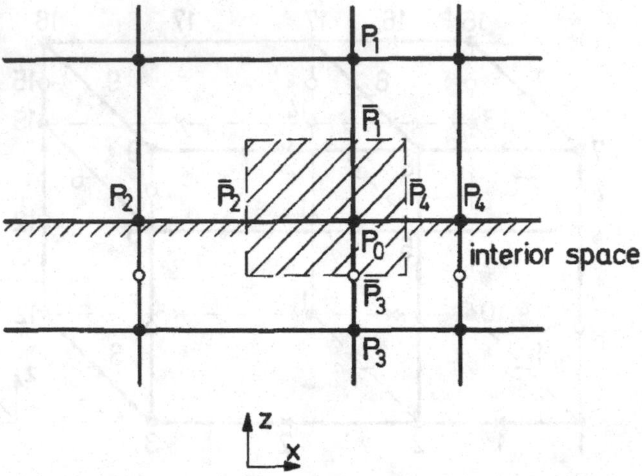

• mesh point

o calculation point of H_z in the interior space

Fig.:3 Nomenclature for the five-point formula (in two dimensions)

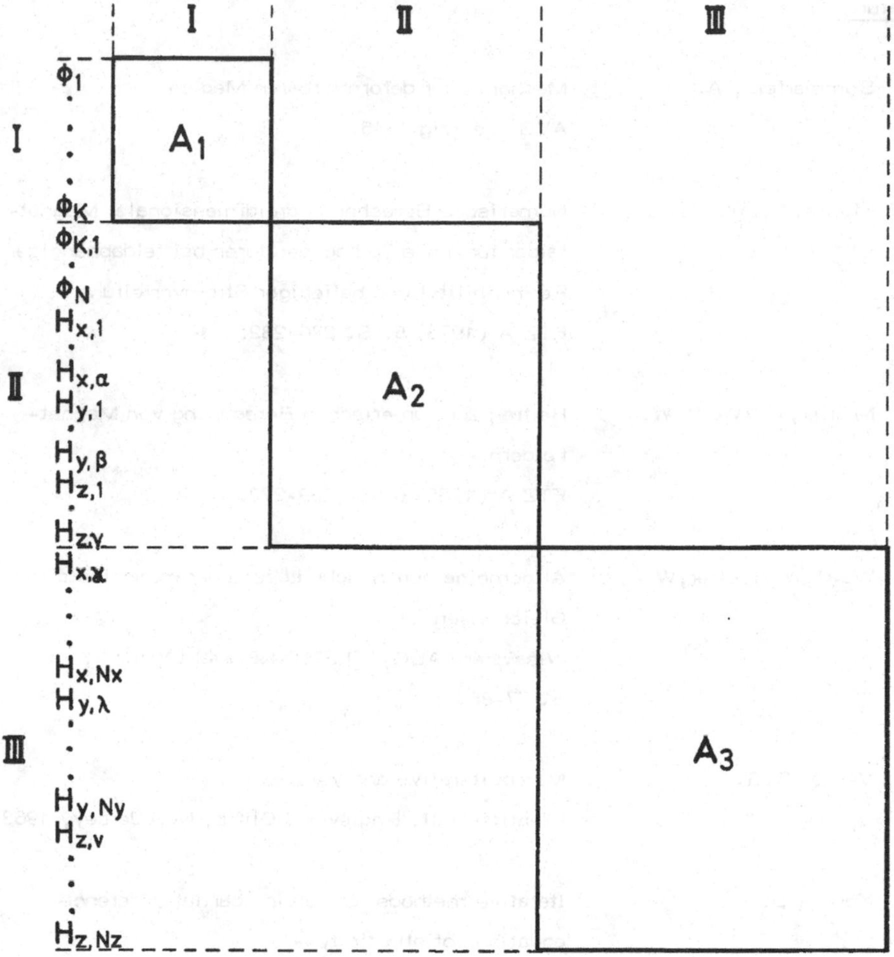

Figure 4 Schematic representation of the coefficient matrix

Literature

/1/ Sommerfeld, A. Mechanik der deformierbaren Medien
AVG, Leipzig 1945.

/2/ Müller, W./Wolff, W. Numerische Berechnung dreidimensionaler Magnetfelder für große Turbogeneratoren bei feldabhängiger Permeabilität und beliebiger Stromverteilung
ETZ A (1973) 5, S. 276-282.

/3/ Müller, W./Wolff, W. Beitrag zur numerischen Berechnung von Magnetfeldern
ETZ A (1975) 6, S. 269-273.

/4/ Wolff, W./Müller, W. Allgemeine numerische Lösung der magnetostatischen Gleichungen
Wiss. Ber. AEG-TELEFUNKEN 49 (1976) 3, S. 77-86.

/5/ Varga, R.S. Matrix iterative analysis
Prentice Hall, Englewood Cliffs, New Jersey, 1963.

/6/ Young, D. Iterative methods for solving partial difference equations of elliptic type
Trans.Amer.Math.Soc. 76, 1954, PP 92-111.

/7/ Young, D. Iterative solution of large linear systems
Academic Press, New York, 1971.

/8/ Kredell, B. On complex successive overrelaxation
BIT 2, 1962, PP 143-152.

BOUNDARY-VALUE TECHNIQUE FOR THE NUMERICAL SOLUTION OF PERIODIC PARABOLIC PROBLEMS

H. Koßmann

1. A Physical Example

We consider the problem of flow of heat in a thin ring-shaped heat conductor whose temperature depends only on the coordinate x and on the time t. The corresponding equation is (see [9])

$$c\rho\frac{\partial u}{\partial t} = \frac{\partial}{\partial x}\left(k\frac{\partial u}{\partial x}\right) + f(x,t) - \frac{hp}{\sigma}u \ , \quad 0 < x < 1 \ , \quad t > 0 \ ,$$

with the initial temperature data

$$u(x,0) = r(x) \ , \quad 0 \leqslant x \leqslant 1 \ ,$$

and the conditions of periodicity

$$u(0,t) = u(1,t) \ , \quad \frac{\partial u(0,t)}{\partial x} = \frac{\partial u(1,t)}{\partial x} \ , \quad t \geqslant 0 \ .$$

Here, u denotes the temperature, which is the unknown function of the variables x and t, k is the thermal conductivity of the ring, c is the specific heat, ρ is the density, and f(x,t) is the strength of heat sources located in the ring. The term $-\frac{hp}{\sigma}u$ takes into consideration the loss of heat through the lateral surface where the ambient temperature is taken to be equal to zero. In this term h denotes the coefficient of external thermal conductivity, p is the perimeter of a cross section of the ring perpendicular to the x axis, and σ is the area of such a cross section.

In this paper we get a numerical approximation for solutions periodic in time by using a coupled system of difference equations like for elliptic problems. This method is called the "boundary-value technique" and has been suggested by Gilmour [4] and Greenspan [5] for the first initial-boundary value problem. Carasso-Parter [1] and Gekeler [3] gave an analysis of the method.

Here we use higher order correct difference equations. Direct methods are used to solve the resulting system of algebraic equations in the linear case, whose number of arithmetic operations is independent of the order of the difference method. For that reason we can get a good approximation with only a few operations needed.

Better stability is achieved if the boundary values are known or if the solution u is odd-periodic, i.e.

$$u(x,t) = -u(x+\frac{1}{2},t) , \quad 0 < x \leq \frac{1}{2} ,$$

and if these properties are employed in the boundary-value technique.

2. Nonlinear Problems

Let G be the interval (0,1] and let $D=G\times(0,\infty)$. We now consider for convenience the initial boundary value problem

$$e(x)u_t = a(t)u_{xx} + b(x,t)u_x - f(x,t,u) , \quad (x,t) \in D ,$$

$$(1) \qquad u(x,0) = r(x) , \quad x \in G ,$$

$$u \text{ periodic in } x \text{ with period } 1 ,$$

satisfying the following conditions:

Assumption

(i) Let the problem be elliptic in space: $a(t) \geq \alpha > 0$, $t \in (0,\infty)$, and let $a \in C(0,\infty)$.

(ii) Further, let b be continuously differentiable in D, $|b_x(x,t)| \leq \gamma$ in D, and let $e \in C[0,1]$.

(iii) We assume that f is continuous in $D\times\mathbb{R}$ and

$$(f(x,t,v)-f(x,t,w))(v-w) \geq \delta(v-w)^2 \;\forall\; (x,t)\in D, \;\forall\; v,w\in\mathbb{R}, \text{ with } \delta\in\mathbb{R}$$

(iv) Let a,b,f be periodic in t with period T and let e,b,f,r be periodic in x with period 1.

(v) Suppose there exists a classical solution u which is periodic in t with period T and let u_t , u_{xx} be continuous in D .

In the following section we consider a numerical approach to the above introduced analytical problem.

3. Discrete Approximation to the Analytic Problem

Because of the periodicity it suffices to consider the rectangle $G\times(0,T]$ for the numerical solution.

Let Δx , Δt be small increments of the variables x,t and introduce a mesh by means of the mesh-points $(k\Delta x, n\Delta t)$, $k,n \in \mathbb{N}$. Let M,N be positive integers which denote the number of mesh-lines in x- and t-direction in $G\times(0,T]$.

We will be dealing with functions defined at the mesh-points and we adopt the notation

$$v_k^n = v(k\Delta x, n\Delta t) \ , \quad f_k^n(p) = f(k\Delta x, n\Delta t, p)$$

(analogous, if k,n are not integers).

Let V^n denote the M-component vector $V^n = (v_1^n, \ldots, v_M^n)^T$ and let V be the "block" vector of MN components

$$V = \begin{bmatrix} V^1 \\ \vdots \\ V^N \end{bmatrix}.$$

We define the following norm and scalar product on $\mathbb{R}^{MN}$

$$(V,W) = \frac{\Delta x}{N} \sum_{n=1}^{N} \sum_{k=1}^{M} v_k^n w_k^n \ , \qquad \| V \|_2 = \sqrt{(V,V)} \ .$$

Now we define difference operators at the mesh-point $(k\Delta x, n\Delta t)$ as follows:

The central differences relative to x

$$\delta_x v_k^n = v_{k+\frac{1}{2}}^n - v_{k-\frac{1}{2}}^n \ , \quad \delta_x^p v_k^n = \delta_x(\delta_x^{p-1} v_k^n) \ , \quad p=1,2,3,\ldots,$$

the forward differences relative to x

$$\Delta_x v_k^n = v_{k+1}^n - v_k^n \ , \quad \Delta_x^p v_k^n = \Delta_x(\Delta_x^{p-1} v_k^n) \ , \quad p=1,2,3,\ldots,$$

and the forward differences relative to t

$$\Delta_t v_k^n = v_k^{n+1} - v_k^n \ , \quad \Delta_t^p v_k^n = \Delta_t(\Delta_t^{p-1} v_k^n) \ , \quad p=1,2,3,\ldots \ .$$

From Stirling's interpolation polynomial we can obtain the following symmetric formulae to approximate $(u_{xx})_k^n$, $(u_x)_k^n$, and $(u_t)_k^n$ respectively:

$$\delta_{x,L}^2 v_k^n = \frac{1}{\Delta x^2} \sum_{j=1}^{L} 2 \frac{(-1)^{j-1}[(j-1)!]^2}{(2j)!} \delta_x^{2j} v_k^n ,$$

$$\Delta_{x,L} v_k^n = \frac{1}{\Delta x} \sum_{j=1}^{L} \frac{(-1)^{j-1}[(j-1)!]^2}{(2j-1)!} \cdot \frac{\Delta_x^{2j-1} v_{k-(j-1)}^n + \Delta_x^{2j-1} v_{k-j}^n}{2} ,$$

$$\Delta_{t,L} v_k^n = \frac{1}{\Delta t} \sum_{j=1}^{L} \frac{(-1)^{j-1}[(j-1)!]^2}{(2j-1)!} \cdot \frac{\Delta_t^{2j-1} v_k^{n-(j-1)} + \Delta_t^{2j-1} v_k^{n-j}}{2} .$$

For $g \in C^{2L+2}$ we have the estimate (see Kantorovich-Krylov [7])

$$\delta_{x,L}^2 g(x) = g''(x) + O(\Delta x^{2L}) ,$$

and for $g \in C^{2L+1}$ we obtain

$$\Delta_{x,L} g(x) = g'(x) + O(\Delta x^{2L}) .$$

Now let $\Delta x = \frac{1}{M}$, $\Delta t = \frac{T}{N+1}$.

Our finite-difference approximation to (1) will be

$$\text{(2)} \quad \begin{aligned} e_k \Delta_{t,L_t} v_k^n &= a^n \delta_{x,L_x}^2 v_k^n + b_k^n \Delta_{x,L_x} v_k^n - f_k^n(v_k^n) , \quad 1 \leq k \leq M , \quad 1 \leq n \leq N , \\ v_k^o &= r_k = v_k^{N+1} , \quad 1 \leq k \leq M . \end{aligned}$$

Because of the periodicity we set $v_{k+M}^n = v_k^n$, $v_k^{n+N+1} = v_k^n$.

We collect the MN equations (2) in the usual way, and we obtain the following system of equations

$$\text{(3)} \qquad P(V) = (S+A+B)V + F(V) = H .$$

Here, A is the (MN×MN)-matrix, which describes the linear operator

$$A : \mathbb{R}^{MN} \ni V \longrightarrow (-a^n \delta_{x,L_x}^2 v_k^n)_{\substack{1 \leq k \leq M \\ 1 \leq n \leq N}} \in \mathbb{R}^{MN} ,$$

B is the (MN×MN)-matrix, which describes the linear operator

$$B : \mathbb{R}^{MN} \ni V \longrightarrow (-b_k^n \Delta_{x,L_x} v_k^n)_{\substack{1\leq k\leq M \\ 1\leq n\leq N}} \in \mathbb{R}^{MN},$$

S is the (MN×MN)-matrix and H the MN-vector, which describe the affine mapping

$$S_a : \mathbb{R}^{MN} \ni V \longrightarrow (e_k \Delta_{t,L_t} v_k^n)_{\substack{1\leq k\leq M \\ 1\leq n\leq N}} = SV - H \in \mathbb{R}^{MN}.$$

Here, H contains only known quantities, i.e. H depends on the initial values

$$v_k^o = r_k = v_k^{N+1}, \quad 1 \leq k \leq M.$$

Further, F(V) is the MN-vector, which contains the elements $f_k^n(v_k^n)$.

Obviously the estimates below are true.

Theorem 1

Let U be the vector obtained from evaluations of u(x,t) at the mesh points. Then we have

$$\|P(U) - H\|_2 \longrightarrow 0 \quad \text{as} \quad \Delta x \longrightarrow 0 \quad \text{and} \quad \Delta t \longrightarrow 0.$$

Now let

$$u \in C^{2L_{max}+2}, \quad L_{max} = \max\{L_x, L_t\}.$$

Then we obtain

$$\|P(U) - H\|_2 \leq K(\Delta t^{2L_t} + \Delta x^{2L_x}),$$

where K is a constant independent of Δx, Δt and T.

4. Stability and Convergence

4.1

We use the monotonicity in the sense of Minty to prove the stability relative to the $\|.\|_2$-norm, i.e. with a suitable constant K we show that the following inequality is true

$$\|U - V\|_2 \leqslant K\|P(U) - P(V)\|_2 .$$

First we consider the contribution of the (MN×MN)-matrix A to the monotonicity. Let H_L be the (M×M)-matrix which describes the linear operator

$$H_L : \mathbb{R}^M \ni V^n \longrightarrow (-\delta^2_{x,L} v^n_k)_{1\leqslant k\leqslant M} \in \mathbb{R}^M \quad (\text{with } v^n_{k+M}=v^n_k) .$$

With $D_a=\operatorname{diag}(a^1, \ldots ,a^N)$ we get $A=D_a \otimes H_{L_x}$.

The eigenvalues of H_L are given by (see [8])

$$(4) \qquad \Lambda_{p,L} = \frac{1}{\Delta x^2} \sum_{j=1}^{L} \frac{2^{2j+1}[(j-1)!]^2}{(2j)!} \sin^{2j}(\frac{p\pi}{M}) , \quad 1 \leqslant p \leqslant M ,$$

with the corresponding orthogonal eigenvectors

$$(5) \qquad \begin{aligned} X_p &= (\sin\frac{2\pi p}{M}, \sin\frac{4\pi p}{M}, \ldots , \sin 2\pi p)^T , && 1 \leqslant p \leqslant \frac{M-1}{2} , \\ X_p &= (\cos\frac{2\pi p}{M}, \cos\frac{4\pi p}{M}, \ldots , \cos 2\pi p)^T , && \frac{M}{2} \leqslant p \leqslant M . \end{aligned}$$

Because of assumption (i) we have $a^n>0$, $1\leqslant n\leqslant N$, and further we have $\Lambda_{p,L_x} > 0$, $\Lambda_{M,L_x} = 0$.

Hence the matrix $A = D_a \otimes H_{L_x}$ is monotone, i.e. positive semidefinite.

Now we consider the contribution of the (MN×MN)-matrix B to the monotonicity. Let

$$\beta_L = \frac{1}{2}\Big(\sum_{l=1}^{L} \frac{l[(l-1)!]^2}{(2l-1)!}\Big) + \sum_{l=1}^{L-1}\Big(l^2 \sum_{j=l+1}^{L} \frac{[(j-1)!]^2}{(j-l)!(j+l)!}\Big) .$$

We have $\beta_1 = \frac{1}{2}$, $\beta_2 = \frac{5}{6}$, $\beta_3 = \frac{11}{10}$, and with suitable constants $\underline{c} > 0$, $\bar{c} > 0$ the following inequalities are true (see [8])

$$\underline{c}\sqrt{L} \leq \beta_L \leq \bar{c}\sqrt{L} \quad \forall\, L \geq L_o \,.$$ For $L_o=1$ we can choose $\bar{c}=4$.

We have $|(V,BV)| \leq \gamma\beta_{L_x}(V,V)$. Thus, B disturbs the monotonicity.

The matrix S is skew-symmetric and hence monotone:
We have $(V,SV)=0$.

Because of assumption (iii) we have further

$$(F(V)-F(W),V-W) \geq \delta(V-W,V-W) \,.$$

Therefore the mapping P is uniformly monotone if the inequality

$$c_1 := \delta - \gamma\beta_{L_x} > 0$$

is true:

$$(P(V)-P(W),V-W) \geq c_1(V-W,V-W) = c_1\|V-W\|_2^2 \,.$$

Hence there exists a unique solution of the finite difference approximation (2) (see [10]).

On using Schwarz's inequality we obtain the inequality

$$\|V - W\|_2 \leq \frac{1}{c_1}\|P(V) - P(W)\|_2 \,.$$

Thus the following theorem is true.

Theorem 2

If $c_1=\delta-\gamma\beta_{L_x} > 0$, a unique solution V of the finite difference approximation (2) exists and we obtain stability relative to the $\|.\|_2$-norm with the constant $1/c_1$.

Let U denote the block vector of the solution of the analytic problem at the mesh points and let u be sufficiently smooth, e.g.

$$u \in C^{2L_{max}+2} \quad \text{with} \quad L_{max} = \max\{L_x, L_t\} \,.$$

Then

$$\|U - V\|_2 \leq K(\Delta t^{2L_t} + \Delta x^{2L_x}) ,$$

where K is a constant independent of T.

As in the following finite-difference approximations we get with $\Delta t = \frac{T}{N}$ the same result, if $u(x,0)=r(x)$ is not given and if (2) is correspondingly modified, i.e. only the difference equations are considered.

4.2

Now we assume that the boundary values are known, i.e.

$$u(0,t) = u(1,t) = s(t) , \quad t \geq 0 .$$

Let $\Delta x=\frac{1}{M+1}$, $\Delta t=\frac{T}{N+1}$. Then our finite-difference approximation will be

$$\text{(2')} \quad \begin{aligned} & e_k \Delta_{t,L_t} v_k^n = a^n \delta^2_{x,L_x} v_k^n + b_k^n \Delta_{x,L_x} v_k^n - f_k^n(v_k^n) , \quad 1 \leq k \leq M , \quad 1 \leq n \leq N , \\ & v_k^0 = r_k = v_k^{N+1} , \quad 1 \leq k \leq M , \quad v_0^n = v_{M+1}^n = s(n\Delta t) , \quad 1 \leq n \leq N . \end{aligned}$$

Because of the periodicity we set $v_{k+M+1}^n = v_k^n$, $v_k^{n+N+1} = v_k^n$.

We collect the MN equations (2') in the usual way, and we obtain the following system of equations

$$\text{(3')} \quad P_R(V) = (S+A_R+B_R)V + F(V) = H_R .$$

H_R contains only known quantities, i.e. H_R depends on the initial values

$$v_k^0 = r_k = v_k^{N+1} , \quad 1 \leq k \leq M ,$$

and on the boundary values

$$v_0^n = v_{M+1}^n = s(n\Delta t) , \quad 1 \leq n \leq N .$$

The influence of S , B_R and F on the monotonicity is analogous as above. We now consider the contribution of the (MN×MN)-matrix A_R to the monotonicity.

Let H_L^R be the (M×M)-matrix which describes the linear part of the affine mapping

$$H^R_{a,L} : \mathbb{R}^M \ni V^n \longrightarrow (-\delta^2_{x,L} v^n_k)_{1\leq k\leq M} = H^R_L V^n + R \in \mathbb{R}^M$$

$$(\text{with } v^n_{k+M+1} = v^n_k).$$

With $D_a = \mathrm{diag}(a^1, \ldots, a^N)$ we get $A_R = D_a \otimes H^R_{L_x}$.

We decompose the matrix H^R_L to get an estimation of its eigenvalues:

$$H^R_L = H^R_1 + (H^R_L - H^R_1) .$$

The eigenvalues of H^R_1 are known (see Polozhii [11]), and there results

$$X^T H^R_1 X \geq 8 X^T X \quad \forall\, X \in \mathbb{R}^M .$$

The matrix $H^R_L - H^R_1$ is monotone, too, (see [8]) and hence we get

$$(A_R V, V) \geq 8\alpha (V,V) .$$

Thus we obtain the following theorem.

<u>Theorem 3</u>

If $c_2 = 8\alpha + \delta - \gamma\beta_{L_x} > 0$, a unique solution V of the finite difference approximation (2') exists and we obtain stability relative to the $\|.\|_2$-norm with the constant $1/c_2$.

Let $u \in C^{2L_{max}+2}$ with $L_{max} = \max\{L_x, L_t\}$. Then

$$\|U - V\|_2 \leq K(\Delta t^{2L_t} + \Delta x^{2L_x}) .$$

Thus better stability can be achieved if the boundary values are known.

<u>4.3</u>

Now we assume that the solution u of (1) is odd-periodic, i.e.

$$u(x,t) = -u(x+\tfrac{1}{2},t) , \quad 0 < x \leq \tfrac{1}{2} .$$

Further, let r be odd-periodic and let b be periodic in x with period $\frac{1}{2}$.

Because of the odd-periodicity it suffices to approximate the solution u at the mesh points in $(0,\frac{1}{2}] \times (0,T]$. With

$$\Delta x = \frac{1}{2M} \quad , \quad \Delta t = \frac{T}{N+1}$$

we choose the same difference equations as in (2), and because of the odd-periodicity we set

$$v^n_{k+M} = -v^n_k .$$

Now the (M×M)-matrix H^u_L , which describes the linear operator

$$H^u_L : \mathbb{R}^M \ni V^n \longrightarrow (-\delta^2_{x,L} v^n_k)_{1\leq k\leq M} \in \mathbb{R}^M \quad (\text{with } v^n_{k+M}=-v^n_k)$$

has the eigenvalues (see [8])

$$\Lambda^u_p = \frac{1}{\Delta x^2} \sum_{j=1}^{L} \frac{2^{2j+1}[(j-1)!]^2}{(2j)!} \sin^{2j}\left(\frac{p\pi}{2M}\right) \quad , \quad 1 \leq p \leq M \text{ , } p \text{ odd ,}$$

with the corresponding orthogonal eigenvectors

$$Y_p = \left(\cos\frac{\pi p}{M}, \cos\frac{2\pi p}{M}, \ldots ,\cos \pi p\right)^T , \quad 1 \leq p \leq M \text{ , } \quad p \text{ odd ,}$$

$$Y_{p+1} = \left(\sin\frac{\pi p}{M}, \sin\frac{2\pi p}{M}, \ldots ,\sin \pi p\right)^T , \quad 1 \leq p < M \text{ , } \quad p \text{ odd .}$$

Because of $\Lambda^u_p \geq 32$ the theorem below is true.

<u>Theorem 4</u>

If

$$c_3 = 32\alpha + \delta - \gamma\beta_{L_x} > 0 \quad ,$$

a unique solution V of the finite difference approximation in the odd periodic case exists and we obtain stability relative to the $\|.\|_2$-norm with the constant $1/c_3$.

Let $u \in C^{2L\max+2}$. Then

$$\|U - V\|_2 \leq K(\Delta t^{2L_t} + \Delta x^{2L_x}) .$$

Thus we get better stability if the solution is odd-periodic.

5. Direct Solution Using Fast Fourier Transforms

Now we consider direct methods to solve the difference approximations in the linear case, and we treat the boundary value problem

$$u_t = au_{xx} - cu + d(x,t) \ , \quad 0 < x \leqslant 1 \ , \quad t > 0 \ ,$$

u periodic in x with period 1 and u periodic in t with period T.

With

$$\Delta x = \frac{1}{M} \ , \quad \Delta t = \frac{T}{N}$$

our finite-difference approximation will be

$$(6) \qquad \Delta_{t,L_t} v_k^n = a\delta^2_{x,L_x} v_k^n - cv_k^n + d_k^n \ , \quad 1 \leqslant k \leqslant M \ , \ 1 \leqslant n \leqslant N \ .$$

Because of the periodicity we set $v^n_{k+M} = v^n_k$, $v^{n+N}_k = v^n_k$.

We use a direct method whose number of arithmetic operations is independent of L_x and L_t for the $O(\Delta x^{2L_x} + \Delta t^{2L_t})$ - approximation.

Let H be the (M×M)-matrix, which describes the linear operator

$$H : \mathbb{R}^M \ni v^n \longrightarrow (-\delta^2_{x,L_x} v_k^n)_{1 \leqslant k \leqslant M} \in \mathbb{R}^M \ ,$$

and let S be the (N×N)-matrix, which describes the linear operator

$$S : \mathbb{R}^N \ni V_k \longrightarrow (\Delta_{t,L_t} v_k^n)_{1 \leqslant n \leqslant N} \in \mathbb{R}^N \ .$$

If we collect the MN equations (6) in the usual way, we obtain the following system of equations with $L = aH + cE_M$

$$(S \otimes E_M + E_N \otimes L)V = H$$

(E_K identity matrix of dimension K).

H has the eigenvalues (4) with the corresponding orthogonal eigenvectors (5).

Let

$$\Lambda = \text{diag}(a\Lambda_{1,L_x}+c, \ldots ,a\Lambda_{M,L_x}+c)$$

and let Q be the (M×M)-matrix

$$Q = (Y_1, \ldots ,Y_M) ,$$

where the Y_j are the orthonormal eigenvectors

$$Y_j = \begin{cases} \sqrt{\frac{2}{M}}\, X_j & , \quad 1 \leqslant j \leqslant \frac{M-1}{2} , \quad \frac{M+1}{2} \leqslant j \leqslant M-1 \\ \sqrt{\frac{1}{M}}\, X_j & , \quad j=M , \text{ and } j = \frac{M}{2} \text{ if } M \text{ is even.} \end{cases}$$

We have

$$Q^T L Q = \Lambda .$$

With the MN-vectors

$$\overline{V} = (E_N \otimes Q^T)V , \quad \overline{H} = (E_N \otimes Q^T)H$$

we find that

$$(E_N \otimes Q^T)(S \otimes E_M + E_N \otimes L)(E_N \otimes Q)(E_N \otimes Q^T) V$$

$$= (S \otimes E_M + E_N \otimes \Lambda)\, \overline{V} = \overline{H} .$$

Let

$$r_s = e^{i\frac{2s\pi}{N}} .$$

The eigenvalues of S are given by (see [8])

$$\eta_s = \frac{i}{\Delta t} \sum_{j=1}^{L_t} \frac{4^{j-1}[(j-1)!]^2}{(2j-1)!} \sin^{2j-2} \frac{s\pi}{N} \sin \frac{2s\pi}{N} , \quad 1 \leqslant s \leqslant N ,$$

with the corresponding orthonormal eigenvectors

$$z^s = \frac{1}{\sqrt{N}}(1,r_s, \ldots ,r_s^{N-1})^T .$$

Let P be the (N×N)-matrix

$$P = (z^1, \dots, z^N)$$

and let

$$\Phi = \text{diag}\,(\eta_1, \dots, \eta_N)\ .$$

Then we have $\bar{P}^T S P = \Phi$.
With the MN-vectors

$$\hat{V} = (\bar{P}^T \otimes E_M)\,\bar{V}\ , \qquad \hat{H} = (\bar{P}^T \otimes E_M)\,\bar{H}$$

we find that

$$(\bar{P}^T \otimes E_M)(S \otimes E_M + E_N \otimes \Lambda)(P \otimes E_M)(\bar{P}^T \otimes E_M)\,\bar{V}$$

$$= (\Phi \otimes E_M + E_N \otimes \Lambda)\,\hat{V} = \hat{H}\ .$$

We obtain

$$\hat{v}_k^n = \frac{1}{\lambda_k + \eta_n}\,\hat{h}_k^n\ .$$

Now we get the solution

$$V = (E_N \otimes Q)(P \otimes E_M)\,\hat{V}\ .$$

If $M = 2^q$, $N = 2^r$, $q,r \in \mathbb{N}$, we can use the fast Fourie transform (see Hockney [6], Cooley-Tuckey [2]) to compute

$$\hat{H} = (\bar{P}^T \otimes E_M)(E_N \otimes Q^T)\,H$$

and

$$V = (E_N \otimes Q)(P \otimes E_M)\,\hat{V}\ .$$

Example

To get some idea of the error of the approximations we consider the following simple problem

$$u_t - u_{xx} + u = \sin 2\pi x\, [\cos t + (1+4\pi^2) \sin t] \quad ,$$

u periodic in x with period 1 and u periodic in t with period 2π, with the solution

$$u(x,t) = \sin 2\pi x \sin t \; .$$

We choose the finite-difference approximation (6) and we obtain the following errors

M	N	L_x	L_t	$\|U - V\|_\infty$
20	20	1	1	$8 \cdot 10^{-3}$
20	20	6	6	$3{,}17 \cdot 10^{-10}$

References

1. Carasso, A., Parter, S.V.: An analysis of "boundary-value techniques" for parabolic problems. Math. Comput. 24, 315-340 (1970)
2. Cooley, J.W., Tuckey, J.W.: An algorithm for the machine calculation of complex Fourier series. Math. Comput. 19, 297-301 (1965)
3. Gekeler, E.: Long-range and periodic solutions of parabolic problems. Math.Z. 134, 53-66 (1973)
4. Gilmour, A.: Solution of certain unsteady heat flow problems by relaxation methods. British J. Appl. Phys. 2, 199-204 (1951)
5. Greenspan, D.: Approximate solution of initial-boundary parabolic problems by boundary value techniques. MRC Technical Report Nr. 782, Madison, Wis. (1967)
6. Hockney, R.W.: A fast direct solution of Poisson's equation using Fourier analysis. Journ. ACM 12, 95-113 (1965)
7. Kantorovich, L.V., Krylov, V.I.: Approximate methods of higher analysis. New York: Interscience Publishers 1964
8. Koßmann, H.: Randwerttechnik zur numerischen Lösung des vierten Randwertproblems für parabolische Differentialgleichungen. Dissertation, Ruhr-Universität Bochum 1977
9. Mikhlin, S.G., ed.: Linear equations of mathematical physics. New York: Holt, Rinehart and Winston 1964
10. Ortega, J.M., Rheinboldt, W.C.: Iterative solution of nonlinear equations in several variables. New York-London: Academic Press 1970
11. Polozhii, G.N.: The method of summary representation for numerical solution of problems of mathematical physics. Oxford: Pergamon Press 1965

TIME-DISCRETISATIONS FOR NONLINEAR EVOLUTION EQUATIONS

by

H.Kreth

Introduction: Many mathematical models for transport processes in fluid-dynamics, as for example the diffusion-transport equation or the Burgers equation, are special cases for time-dependent and/or nonlinear evolution equations in the sense of Kato [3] . The difference between the solution of the evolution equation and the solution for disturbed initial data and the disturbed differential equation can be estimated by a stability inequality. Discrete methods for evolution equations should not only have high accuracy, but the solution of the discrete problem should have the same properties as the solution of the differential equation. This means specially, that for time-steps $h \to 0$ the error estimation for the difference method should have the same time-dependence as the stability inequality; we call this a "discrete analogon". Usually in the literature such error estimations for difference methods are proofed only for the linear test-equation $y'=\lambda y$ with $Re\lambda < 0$ (see f.e.[5]), and it is not clear, if assumptions like A-stability and consistency are sufficient to get error estimations with accurate time-dependence in the case of nonlinear problems. In recent years Dahlquist [1] and Nevanlinna [8], [9] proved global error estimations for autonomous evolution equations, but the estimations are accurate in time for the implicit Euler-method only. In this paper we look at the second-order backward-difference method and prove an estimation, which is a discrete analogon to the stability inequality for time-dependent and/or nonlinear evolution equations.

Multistep time-discretisations for evolution equations.

Let H be a (real or complex) Hilbert-space with $\|u\| = (u,u)^{1/2}$ and $[0,T]$ a real interval. Let $\{F(t)\}$ be a family of continous operators mapping H into itself and monotone in the sense of Minty [7], that means there exists a continous function $b : [0,T] \to [0,\infty)$ with

$$Re(F(t)u-F(t)v,u-v) \geq b(t) \|u-v\|^2 \qquad (1)$$

for all $u,v \in H$ and $t \in [0,T]$. Additionally there may exist a constant L, such that

$$\|F(t)v-F(s)v\| \leq L|t-s|(1+\|v\|+\|F(s)v\|)$$

for all $v \in H$ and $s,t \in [0,T]$. Under these assumptions the initial-value problem

$$u_t + F(t)u = 0 \ , \ t \in [0,T] \ , \qquad u(0) = u_o \ , \tag{2}$$

has a unique solution $u(t)$ for all $u_o \in H$ (see [3]).
If $v(t)$ is the solution of the disturbed problem

$$v_t + F(t)v = e(t) \ , \ t \in [0,T] \ , \qquad v(0) = u_o + e_o \ , \tag{3}$$

where $e(t)$ is a lipschitzcontinous mapping from $[0,T]$ into H, then $\|u(t)-v(t)\|$ can be estimated by the stability inequality

$$\|u(t)-v(t)\| \leq \exp(-\int_0^t b(s)ds)[\|e_o\| + \int_0^t \|e(s)\| \exp(\int_0^s b(r)dr)ds]. \tag{4}$$

With $e(t) \equiv 0$ it follows, that the solution of (2) becomes independent of the initial data for great T, if

$$\int_0^t b(s)ds \to \infty \quad \text{for} \quad t \to \infty.$$

Remark: Normally in the theory of evolution equations the family $\{F(t)\}$ is only defined on a subset $D \subset H$ and is in general not continous, but maximal monotone (see [3]). In this paper we are only interested in time-discretisations and assume, that we have constructed a so-called "semidiscrete initial-value problem" of the form (2) by Galerkin-methods or difference methods in space direction. Under appropriate propositions estimations between the solution of the non-regular evolution equation and the solution of the semidiscrete problem are possible (see f.e. [2]).

To solve the initial-value problem (2) we use multistep difference methods of the form

$$\sum_{j=0}^{k} \alpha_j u_{n+j} + hF(\sum_{j=0}^{k} \beta_j t_{n+j})(\sum_{j=0}^{k} \beta_j u_{n+j}) = 0 \tag{5}$$

with $\beta_k > 0$, $\alpha_k > 0$, $\sum_{j=0}^{k} \beta_j = 1$ (see [1]). If $u_n, \ldots, u_{n+k-1}$ is known, u_{n+k} is uniquely determinated by (5).

If one is specially interested to solve problems of fluid dynamics, the difference method (5) should have the following properties:

1. If H is semi-ordered and the solution $u(t)$ of (2) is positive, then the numerical solution u_n of (5) should be positive. There should be no numerical oscillations.

2. The solution operator of (5) should mollify in the same way as the solution operator of (2), that means the method (5) should only have small numerical damping properties. (A test-equation for 1. and 2. is the equation $u_t + au_x = bu_{xx}$ with $a \gg b > 0$.)
3. Often T is very great and so there should be no restrictions for the stepsize h.
4. For the method (5) there should exist a error estimation, which is a discrete analogon to the stability inequality (4).

The properties 1.-4. are severe restrictions for the class of multistep methods. If we consider 3. and 4., we have to use A-stable methods only. The property 2. can be achieved by high accuracy and a possibility to fulfil 1. is to avoid centered differences. This means specially not to use the Crank-Nicolson method for time-discretisation. Summarizing we have to look at two-step second-order A-stable methods, if we want to minimize k. This class of methods was characterized by Liniger [6] and the most famous one is the second-order backward-difference method

$$\frac{3}{2} u_{n+2} - 2u_{n+1} + \frac{1}{2} u_n + hF(t_{n+2})u_{n+2} = 0 \quad . \tag{6}$$

For the second-order BD-method we now proof the property 4.

<u>Error estimation for the second-order BD-method.</u>

For known u_0, u_1, v_0, v_1 two sequences $\{u_n\}$ and $\{v_n\}$ are defined by

$$\begin{aligned} \frac{3}{2} u_{n+2} - 2u_{n+1} + \frac{1}{2} u_n + hF(t_{n+2})u_{n+2} &= he^1_{n+2} \quad , \\ \frac{3}{2} v_{n+2} - 2v_{n+1} + \frac{1}{2} v_n + hF(t_{n+2})v_{n+2} &= he^2_{n+2} \quad . \end{aligned} \tag{7}$$

To proof a error estimation for the method (6) means to derive a upper bound for $\|u_n - v_n\|$. If $u_n = u(t_n)$ is the solution of (2), then he^1_{n+2} is the local error of the time-discretisation. If v_n is the computed solution, then he^2_{n+2} describes the rounding and iteration errors of the method. For shortness let $w_n = u_n - v_n$ and $e_n = e^1_n - e^2_n$.

The estimations for $\|u_n - v_n\|$ in [1], [8], [9] were evaluated with the theory of quadratic forms. Using this technique the special structure of the method was not fully considered, so that no discrete analogon could be achieved. Further on the function b(t) must be replaced by a lower bound b in [0,T]. For many problems in fluid dynamics b is very small or even b(t) is not strictly positive, so that the estimations are not realistic. The following theorem shows, that one can proof directly a recursion formula for the global error and so get a error estimation with accurate time-dependence.

Theorem : Let (1) be valid and let $\{u_n\},\{v_n\}$ be the sequences defined by (7). Then for all stepsizes $h>0$ and all $n\in\mathbb{N}$ with $2\leq n\leq\frac{T}{h}$ the sequence $\{\|w_n\|\}$ fulfils the inequality

$$\sqrt{\|w_n\|^2+\frac{1}{1+2hb_n}\|2w_n-w_{n-1}\|^2}\leq \exp(h\sum_{j=1}^{n-1}a_j)\sqrt{\|w_1\|^2+\frac{1}{1+2hb_1}\|2w_1-w_0\|^2}$$
$$+4h\sum_{i=1}^{n-1}\exp(h\sum_{j=i}^{n-1}a_j)\|e_{i+1}\| \tag{8}$$

with

$$a_j:=b_j-\frac{\ln(1+2hK)}{hK}\min(b_j,b_{j+1}),$$

where K is a upper bound for b(t) in [0,T].

Proof: For the difference of the two equations (7) we have

$$\frac{3}{2}w_{n+2}-2w_{n+1}+\frac{1}{2}w_n+h(F(t_{n+2})u_{n+2}-F(t_{n+2})v_{n+2})=he_{n+2}.$$

Multiplying with $(1+2hb_{n+2})^2$ and using the Cauchy-Schwarz inequality the scalar product with w_{n+2} leads to

$$(1+2hb_{n+2})^2\operatorname{Re}(3w_{n+2}-4w_{n+1}+w_n,w_{n+2})+2hb_{n+2}(1+2hb_{n+2})^2\|w_{n+2}\|^2$$
$$\leq 2h(1+2hb_{n+2})^2\|e_{n+2}\|\,\|w_{n+2}\|.$$

This inequality becomes symmetrical by additional terms:

$$(1+2hb_{n+2})^2\operatorname{Re}(3w_{n+2}-4w_{n+1}+w_n,w_{n+2})+2hb_{n+2}(1+2hb_{n+2})^2\|w_{n+2}\|^2$$
$$+(1+2hb_{n+2})\operatorname{Re}(3w_{n+1}-4w_n,w_{n+1})+2hb_{n+2}(1+2hb_{n+2})\|w_{n+1}\|^2$$
$$+3\|w_n\|^2+2hb_{n+2}\|w_n\|^2$$
$$\leq(1+2hb_{n+2})\operatorname{Re}(3w_{n+1}-4w_n,w_{n+1})+2hb_{n+2}(1+2hb_{n+2})\|w_{n+1}\|^2$$
$$+3\|w_n\|^2+2hb_{n+2}\|w_n\|^2+2h(1+2hb_{n+2})^2\|e_{n+2}\|\,\|w_{n+2}\|$$

The left side of the inequality is now equivalent to

$$\frac{(1+2hb_{n+2})^3}{2}\|w_{n+2}\|^2 + (1+2hb_{n+2})^2\|\sqrt{2}\,w_{n+2}-\frac{1}{\sqrt{2}}w_{n+1}\|^2$$

$$+ (1+2hb_{n+2})\|\frac{1+2hb_{n+2}}{\sqrt{2}}w_{n+2}-\sqrt{2}\,w_{n+1}+\frac{1}{\sqrt{2}}w_n\|^2$$

$$+\|\frac{1+2hb_{n+2}}{\sqrt{2}}w_{n+1}-\sqrt{2}\,w_n\|^2 + \frac{1+2hb_{n+2}}{2}\|w_n\|^2 ,$$

and if we omit the second term, we get the inequality

$$\frac{(1+2hb_{n+2})^3}{2}\|w_{n+2}\|^2 + (1+2hb_{n+2})^2\|\sqrt{2}\,w_{n+2}-\frac{1}{\sqrt{2}}w_{n+1}\|^2$$

$$\leqslant (1+2hb_{n+2})\mathrm{Re}(3w_{n+1}-4w_n,w_{n+1}) + 2hb_{n+2}(1+2hb_{n+2})\|w_{n+1}\|^2$$

$$+ 3\|w_n\|^2 + 2hb_{n+2}\|w_n\|^2$$

$$-\|\frac{1+2hb_{n+2}}{\sqrt{2}}w_{n+1}-\sqrt{2}\,w_n\|^2 - \frac{1+2hb_{n+2}}{2}\|w_n\|^2$$

$$+ 2h(1+2hb_{n+2})^2\|e_{n+2}\|\,\|w_{n+2}\|$$

$$= \frac{(1+2hb_{n+2})^2}{2}\|w_{n+1}\|^2 + (1+2hb_{n+2})\|\sqrt{2}\,w_{n+1}-\frac{1}{\sqrt{2}}w_n\|^2$$

$$+ 2h(1+2hb_{n+2})^2\|e_{n+2}\|\,\|w_{n+2}\| .$$

So we have proofed the following estimation:

$$\|w_{n+2}\|^2 + \frac{1}{1+2hb_{n+2}}\|2w_{n+2}-w_{n+1}\|^2$$

$$\leqslant \frac{1}{1+2hb_{n+2}}(\|w_{n+1}\|^2 + \frac{1}{1+2hb_{n+2}}\|2w_{n+1}-w_n\|^2)$$

$$+ \frac{4h}{1+2hb_{n+2}}\|e_{n+2}\|\,\|w_{n+2}\| .$$

For time-independent $b(t)\equiv b$ this inequality directly leads to a iterative formula for the global error, for time-dependent $b(t)$ we have to estimate by

$$\|w_{n+2}\|^2 + \frac{1}{1+2hb_{n+2}}\|2w_{n+2}-w_{n+1}\|^2$$

$$\leqslant \frac{1+2hb_{n+1}}{(1+2h\min(b_{n+1},b_{n+2}))^2}(\|w_{n+1}\|^2 + \frac{1}{1+2hb_{n+1}}\|2w_{n+1}-w_n\|^2)$$

$$+ \frac{4h}{1+2hb_{n+2}}\|e_{n+2}\|\sqrt{\|w_{n+2}\|^2 + \frac{1}{1+2hb_{n+2}}\|2w_{n+2}-w_{n+1}\|^2}$$

to get the recursion formula

$$\sqrt{\|w_{n+2}\|^2 + \frac{1}{1+2hb_{n+2}}\|2w_{n+2}-w_{n+1}\|^2}$$

$$\leqslant \frac{\sqrt{1+2hb_{n+1}}}{1+2h\min(b_{n+1},b_{n+2})}\sqrt{\|w_{n+1}\|^2 + \frac{1}{1+2hb_{n+1}}\|2w_{n+1}-w_n\|^2}$$

$$+ \frac{4h}{1+2hb_{n+2}}\|e_{n+2}\|$$

$$\leqslant \exp(ha_{n+1})\left[\sqrt{\|w_{n+1}\|^2 + \frac{1}{1+2hb_{n+1}}\|2w_{n+1}-w_n\|^2} + 4h\|e_{n+2}\|\right] \qquad (9)$$

with

$$a_{n+1} = b_{n+1} - \frac{\ln(1+2hK)}{hK}\min(b_{n+1},b_{n+2}) ,$$

where K is a upper bound for b(t) in [0,T]. The inequality (8) follows by induction. q.e.d.

For $h \to 0$, $n \to \infty$ and $nh \to t \in [0,T]$ we have

$$a_n \to -b(t) \quad \text{and} \quad h\sum_{j=1}^{n-1} a_j \to -\int_0^t b(s)ds ,$$

so that the error estimation (8) for the second-order BD-method is a discrete analogon to the stability inequality (4).

Final remarks:

To achieve recursion formulae of the form (9) is not only possible for the second-order BD-method, but for a class of A-stable 2-step methods of order 2. The proof is highly technical, but similar to the proof of the theorem. Estimations of the form (8) are numerically requested, because for time-discretisations of (2) they ensure, that the numerical solution becomes independent of the initial data for great times in the same way as the solution of (2), and that rounding and iteration errors are damped for increasing time.

In [4] the method (6) was applied to a mathematical model problem for diffusion and transport phenomena in a tidal stream. For this actual problem we have

$$b(t) \geqslant 0, \qquad \int_0^t b(s)ds \to \infty \quad \text{for} \quad t \to \infty ,$$

but not $b(t) \geqslant b > 0$.

The numerical results show, that for this problem the method (6) fulfils the properties 1. and 2. in a sufficient way.

References:

[1] Dahlquist, G. : Error analysis for a class of methods for stiff non-linear initial value problems. Lecture Notes in Math.506, 60-72 (1976)

[2] Helfrich, H.-P. : Fehlerabschätzungen für das Galerkinverfahren zur Lösung von Evolutionsgleichungen. manuscripta math.13, 219-235 (1974)

[3] Kato, T. : Nonlinear semigroups and evolution equations. J.Math.Soc.Japan 19, 508-520 (1967)

[4] Kreth, H. : Ein Zwei-Schritt-Differenzenverfahren zur Berechnung strömungsabhängiger Ausbreitungsvorgänge. to appear in ZAMP 29 (1978)

[5] Lambert, J.-D. : Computational methods in ordinary differential equations. London-New York-Sidney-Toronto:Wiley and Sons 1973

[6] Liniger, W. : A criterion for A-stability of linear multistep integration formulae. Computing 3, 280-285 (1968)

[7] Minty, G.-J. : Monotone (nonlinear) operators in hilbert space. Duke Math.J.29, 341-346 (1962)

[8] Nevanlinna, O. : On error bounds for G-stable methods. BIT 16, 79-84 (1976)

[9] Nevanlinna, O. : On the numerical integration of nonlinear initial value problems by linear multistep methods. BIT 17, 58-71 (1977)

FREQUENCY FITTING IN THE NUMERICAL SOLUTION OF ORDINARY DIFFERENTIAL EQUATIONS

J. D. Lambert

Abstract. The well-known technique of exponential fitting consists of choosing free parameters in a numerical method for a system of ordinary differential equations in such a way that the method gives the exact solution when applied to the scalar test equation $y' = \lambda y$, λ real. This paper considers the extension of this idea to the case of the test equation $y' = Ay$, $y \in \mathbb{R}^2$, A a real 2×2 matrix with eigenvalues $\lambda \pm i\mu$, λ,μ real, with particular reference to the case when $\lambda = 0$, for which the term "frequency fitting" is appropriate. Only one-step methods are considered here.

§1 Introduction

Consider the initial value problem

$$y' = f(y) , \quad y(a) = \eta , \quad y,f \in \mathbb{R}^m \tag{1}$$

and a general one-step method

$$y_{n+1} - y_n = h\phi_f(y_{n+1},y_n;h) , \quad y_0 = \eta \tag{2}$$

which generates the sequence $\{y_n | n=0,1,2,---\}$ where y_n is an approximation to $y(x_n)$ and $x_n = a+nh$. We make the assumption, which is satisfied for all commonly used methods of class (2), that when (2) is applied to the test problem

$$y' = Ay , \quad y(a) = \eta \tag{3}$$

A an $m\times m$ constant matrix, it yields the difference equation

$$y_{n+1} = R(hA)y_n , \quad y_0 = \eta \tag{4}$$

where $R(hA)$ is a rational approximation to $\exp(hA)$. The exact solution to (3) at $x_n = a+nh$ is

$$y(x_n) = [\exp(hA)]^n\eta \tag{5}$$

while the approximate solution given by (2) is

$$y_n = [R(hA)]^n\eta . \tag{6}$$

Let $q \in \mathbb{C}$, and let $R_T^S(q)$ denote an (S,T) rational approximation to $\exp(q)$; if this approximation has maximal order $S+T$, it is the (S,T) Padé approximation which we shall denote by $\hat{R}_T^S(q)$.

Examples

Euler's Rule $\quad y_{n+1} - y_n = hf_n$

$$R(hA) = \hat{R}_0^1(hA) := I + hA$$

Backward Euler Rule $\quad y_{n+1} - y_n = hf_{n+1}$

$$R(hA) = \hat{R}_1^0(hA) := (I - hA)^{-1}$$

Trapezoidal Rule $\quad y_{n+1} - y_n = \frac{1}{2}h(f_{n+1} + f_n)$

$$R(hA) = \hat{R}_1^1(hA) := (I - \tfrac{1}{2}hA)^{-1}(I + \tfrac{1}{2}hA)$$

Let us first review the situation for the familiar case where A has distinct real eigenvalues. Since there will exist a diagonalizing transformation which uncouples the system, it is enough to consider the case $y \in \mathbb{R}^1$, $A = \lambda$, real scalar We ask how well - not merely how stably - the exponential $\exp(h\lambda)$ is represented by $R(h\lambda)$. Some numerical results are given in Table 1.

TABLE 1

$h\lambda$	-.2	-.4	-.6	-.8	-1.2	-1.6	-2.0	-10.0
$e^{h\lambda}$	.819	.670	.549	.449	.301	.202	.135	.000
$\lvert\hat{R}_0^1\rvert$	.8	.6	.4	.2	.2	.6	1.0	9.0
$\lvert\hat{R}_1^0\rvert$	.833	.714	.625	.556	.455	.385	.333	.091
$\lvert\hat{R}_1^1\rvert$	.818	.667	.538	.429	.250	.111	0	.667

For moderately large values of $|h\lambda|$, the representations for $\exp(h\lambda)$ are not particularly good. Our attention is inexorably drawn to the ideas of A- and L-acceptability (Ehle [1]), and to the corresponding A- and L-stability of the method. We now ask what, if anything, is the analogue for the case of a purely imaginary eigenvalue of A (or, more generally, for a complex conjugate pair of eigenvalues of A). An appropriate form of the test problem is

$$y' = Ay, \quad y(a) = \eta, \quad y \in \mathbb{R}^2, \tag{7}$$

where A is a real 2×2 matrix with eigenvalues $\lambda \pm i\mu$, $\mu \neq 0$. There will exist a nonsingular matrix S such that

$$S^{-1}AS = B = \begin{bmatrix} \lambda & -\mu \\ \mu & \lambda \end{bmatrix},$$

and it is straightforward to show that the exact solution (5) of the test problem can be written as

$$y(x_n) = S^{-1}[\exp(h\lambda)]^n \begin{bmatrix} \cos nh\mu & -\sin nh\mu \\ \sin nh\mu & \cos nh\mu \end{bmatrix} S\eta \quad , \tag{8}$$

while the approximate solution (6) may be written

$$y_n = S^{-1} r^n \begin{bmatrix} \cos n\phi & -\sin n\phi \\ \sin n\phi & \cos n\phi \end{bmatrix} S\eta \tag{9}$$

where

$$r \exp(i\phi) = R(h(\lambda+i\mu)) \tag{10}$$

Let us consider the case $\lambda = 0$ and ask how well the exact frequency $h\mu$ is represented by the approximate frequency $\arg R(h\mu)$. Table 2 gives some numerical results.

TABLE 2

$h\mu$	.2	.4	.6	.8	1.2	1.6	2.0	10
$\arg \hat{R}^0_1$, $\arg \hat{R}^1_0$	.197	.381	.540	.675	.876	1.102	1.107	1.471
$\arg \hat{R}^1_1$	.199	.395	.583	.761	1.080	1.349	1.571	2.940
$2\pi/h\mu$	31	16	11	8	5	4	3	-

It is tempting to look for some sort of duality between the ideas leading to Tables 1 and 2, but, as the following points demonstrate, there are fundamental differences between the two ideas.

(1) $\hat{R}^1_0$ and $\hat{R}^0_1$ give very different representation of the damping in Table 1, but give identical representations of the frequency in Table 2.

(2) At a superficial glance, the representations of frequency given in Table 1 seem reasonably good compared with the representations of damping given in Table 1. However, getting an inaccurate representation of damping may not be nearly so serious as getting an inaccurate representation of frequency. Thus:

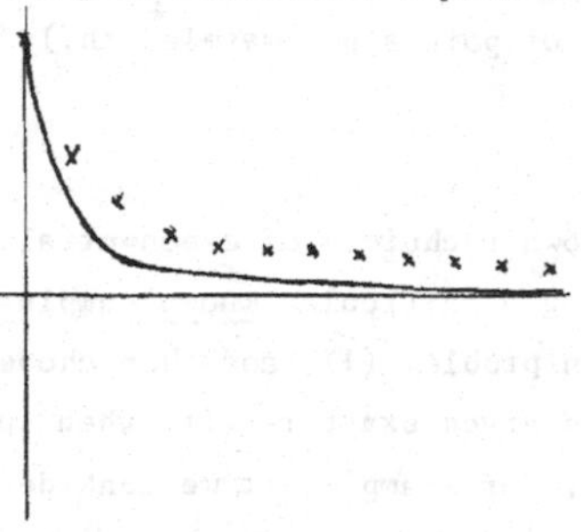

(a)

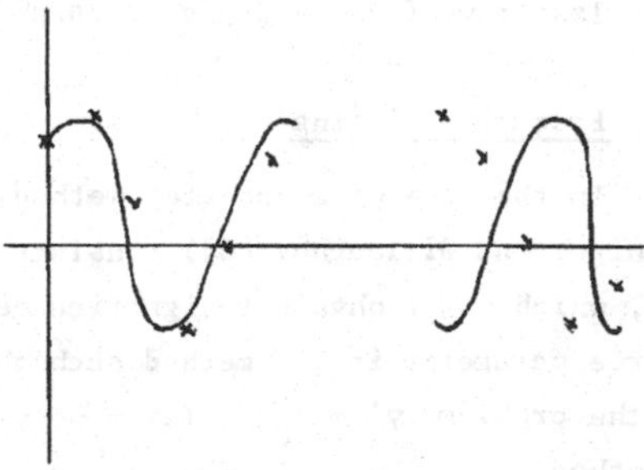

(b)

In (a) the numerical solution asymptotically approaches the exact solution; in (b) it does not.

(3) It is quite wrong to attempt to conceive of "frequency stiffness" as a dual to "stiffness". Stiffness requires us to find stable, but not necessarily accurate, representations for $\exp(h\lambda)$ when $h\lambda \ll 0$. When attempting to represent oscillations, not only are we much more concerned, as in (2) above, with accuracy, but for large values of $h\mu$, the numerical solution will give no representation of the oscillation at all! Thus:

(a)

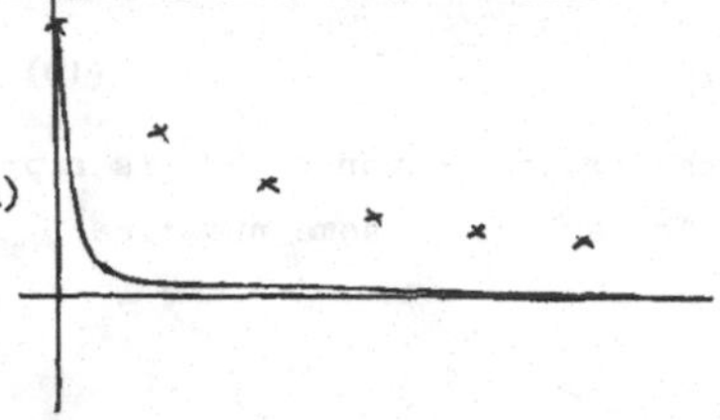

$h\lambda \ll 0$

(b)

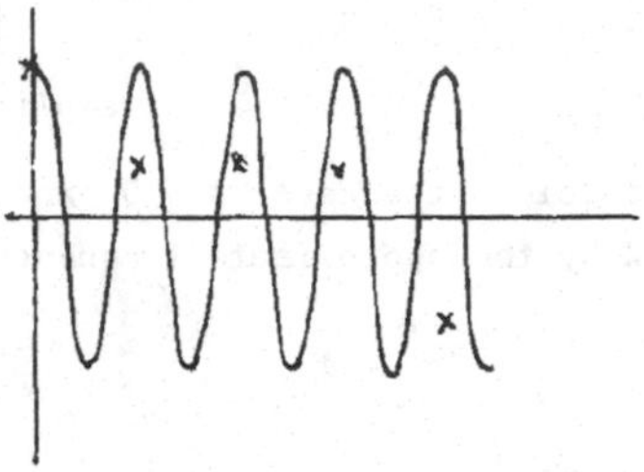

$h\mu \gg 0$

In (a) the numerical solution at least gives the correct qualitative behaviour; in (b) it does not, and may even fail to indicate the existence of any oscillation. Thus, in representing an oscillation, h has to be such that $h\mu$ is not large if we are to have any representation at all! Three or less points per wavelength may not indicate the presence of an oscillation, whereas four or more will:

The wavelength of the exact solution is $2\pi/\mu$ and the number of points per wavelength is $2\pi/h\mu$. Thus, by "not large" in the above underlined statement we mean $2\pi/h\mu \geq 4$ or

$$h\mu \leq \pi/2 \approx 1.6. \tag{11}$$

(The last row of Table 2 gives $2\pi/\mu h$, the number of points per wavelength.)

§2. Frequency Fitting

In the case of a one-step method, the well-known technique of exponential fitting (Liniger and Willoughby [3]) consists of identifying a particular known damping rate, λ_0, which is of physical significance in the given problem (1), and then choosing a free parameter in the method such that the method gives exact results when applied to the problem $y' = \lambda_0 y$, $y(a) = \eta$, $y \in \mathbb{R}^1$. Thus, for example, if we consider the θ-method,

$$y_{n+1} - y_n = h[(1-\theta)f_{n+1} + \theta f_n] ,$$

then equation (4) holds with

$$R(hA) = R(hA,\theta) = [I - (1-\theta)hA]^{-1}[I+\theta hA] \text{ .}$$

Particularising to the case $y \in \mathbb{R}^1$, $A = \lambda_0$, we get that (5) and (6) are identical if we choose

$$\theta = -\frac{1}{h\lambda_0} - \frac{\exp(h\lambda_0)}{1-\exp(h\lambda_0)} \tag{12}$$

Note that we can exponentially fit to only one eigenvalue λ_0, and this must be real; otherwise the method has complex coefficients.

Let us now attempt to extend this idea by replacing the test problem $y' = \lambda y$, $y(a) = \eta$, $y \in \mathbb{R}^1$ by (7) with $\lambda \pm i\mu$ replaced by $\lambda_0 \pm i\mu_0$. It follows from (8), (9) and (10) that we have

<u>exact damping</u> if $\exp(h\lambda_0) = r = |R(h(\lambda_0+i\mu_0))|$,

<u>exact frequency</u> if $h\mu_0 = \phi = \arg R(h(\lambda_0+i\mu_0))$

Note that in (8), damping is a function of λ only, frequency a function of μ only, whereas in (9) damping and frequency are both functions of λ <u>and</u> μ. Thus, for example, if we wish to exponentially fit (in the Liniger-Willoughby sense) to a given complex eigenvalue it is <u>not</u> enough to apply (12) to the real part of the eigenvalue.

In general we will not be able to fit both damping and frequency. It is arguable that practical problems can arise where it is more desirable to fit to a known frequency. (Note that this frequency must arise from an eigenvalue of the Jacobian, and not from an inhomogeneous forcing term such as in $y' = A(x)y + \cos wx$.) In the sequel we shall, except where otherwise stated, assume $\lambda_0 = 0$, that is, that the particular oscillation we fit to is undamped. Analogous, but somewhat heavy, results are easily obtained for the case $\lambda_0 \neq 0$. It now appears to be a simple exercise to choose free parameters in $R(hA) = R_T^S(hA,\alpha,\beta,\text{---})$ to achieve frequency fitting. It turns out, however, that there are snags. It is necessary to consider only cases where $S+T \geq 2$; otherwise, since there will be at least one free parameter, the order of the resulting approximation would be zero, corresponding to the method being inconsistent.

<u>CASE 1</u> $R(hA) = R_0^2(hA, \alpha) := I + hA + \alpha h^2A^2$.

The order is 1 if $\alpha \neq \frac{1}{2}$, 2 if $\alpha = \frac{1}{2}$.

Frequency fitting, that is, $h\mu_0 = \arg R(ih\mu_0)$, is achieved if

$$\alpha = (1-h\mu_0 \cot h\mu_0)/h^2\mu_0^2 \text{ .} \tag{13}$$

However, if this result is to be of any practical value, we cannot ignore stability.

Since $R_T^S(hA)$ cannot be A-acceptable (or even A_0-acceptable (Cryer [2])) when $S > T$, it follows that the corresponding method must have a finite region $\mathcal{R}$ of absolute stability; the numerical solution will be unstable unless the product of h and <u>any</u> eigenvalue of the Jacobian lies in $\mathcal{R}$. The region $\mathcal{R}$ is however a function of α and thus depends on $h\mu_0$ in a complicated manner. However, we can easily check whether the particular product $ih\mu_0$ lies in $\mathcal{R}$, that is whether $r = |R(ih\mu_0)| < 1$. We easily find that, with α given by (13) $r = |h\mu_0 \operatorname{cosec} h\mu_0| > 1 \ \forall \ h\mu_0$. Hence for $R(hA) = R_0^2(hA,\alpha)$, frequency fitting always implies absolute instability; we therefore abandon this case. One wonders whether the above phenomena will occur for $R(hA) = R_0^S(hA,\alpha)$, $S \geq 2$, that is for all explicit methods. This motivates the next case to be considered.

<u>CASE 2</u> $\quad R(hA) = R_0^3(hA,\alpha) := I + hA + \frac{1}{2}h^2A^2 + \alpha h^3A^3$.

The order is 2 if $\alpha \neq \frac{1}{6}$, 3 if $\alpha = \frac{1}{6}$.

Frequency fitting is achieved if

$$\alpha = [h\mu_0 - (1-\tfrac{1}{2}h^2\mu_0^2)\tan h\mu_0]/h^3\mu_0^3$$

With this value of α, we find

$$r = |R(ih\mu_0)| = |(1-\tfrac{1}{2}h^2\mu_0^2)\sec h\mu_0|$$

$$< 1 \quad \text{for} \quad h\mu_0 \in [0,1.48) .$$

Bearing in mind the restriction (11), we see that, for a very reasonable range of $h\mu_0$, the phenomena described in case 1 does not arise. However, we still have the formidable problem of deciding whether the product of h and any <u>other</u> eigenvalue of the Jacobian lies within the region of absolute stability. We are thus motivated to consider cases where $R(hA)$ can be A-acceptable.

<u>CASE 3</u> $\quad R(hA) = R_1^1(hA,\alpha) := [I - \frac{1}{2}(1+\alpha)hA]^{-1}[I + \frac{1}{2}(1-\alpha)hA]$

The order is 1 if $\alpha \neq 0$, 2 if $\alpha = 0$.

A-acceptable iff $\alpha \geq 0$.

($\alpha = 1-2\theta$ corresponds to the θ-method).

Frequency fitting is achieved if

$$\alpha^2 = M(h\mu_0)/h^2\mu_0^2$$

where

$$M(h\mu_0) = h^2\mu_0^2 + 4h\mu_0 \cot h\mu_0 - 4 \tag{14}$$

Clearly, we can frequency fit only when $M(h\mu_0)$ is non-negative; it turns out that this is the case only when $h\mu_0$ lies in one of the intervals

$$[\pi,5.6),\ (2\pi,9), \text{---} (k\pi,(k+1-\epsilon_k)\pi), \text{---} \quad , \quad 0 < \epsilon_k < 1$$

In particular, frequency fitting cannot be achieved for $h\mu_0 < \pi$ and, by the restriction (11), this is unacceptable. The situation is slightly better if we consider the case $\lambda_0 \neq 0$, that is, the case when the significant frequency is damped or attenuated. In the following diagram, frequency fitting is not possible in the shaded regions of the complex $h(\lambda_0+i\mu_0)$-plane

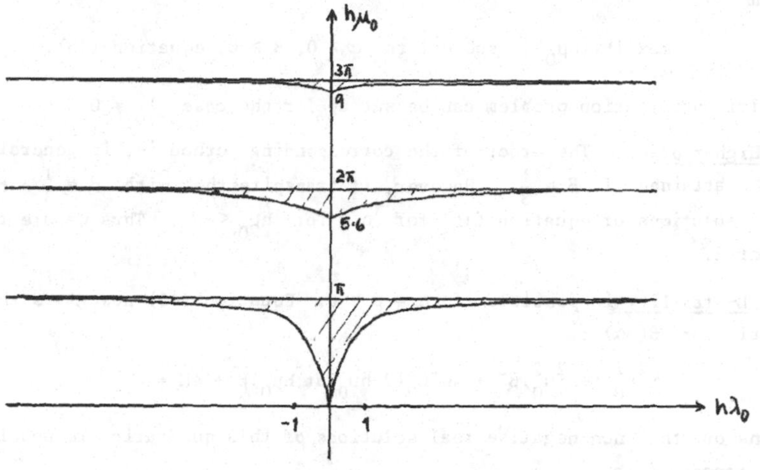

Thus, sufficiently damped or attenuated frequencies can be frequency fitted without contradicting the restriction (11).

<u>CASE 4</u> $R(hA) = R_2^2(hA,\alpha,\beta)$

$$:= [I - \tfrac{1}{2}(1+\alpha)hA + \tfrac{1}{4}(\beta+\alpha)h^2A^2]^{-1}[I + \tfrac{1}{2}(1-\alpha)hA + \tfrac{1}{4}(\beta-\alpha)h^2A^2]$$

The order is 2 if $\beta \neq \frac{1}{3}$, 3 if $\beta = \frac{1}{3}, \alpha \neq 0$, 4 if $\beta = \frac{1}{3}, \alpha = 0$.

A-acceptable iff $\alpha \geq 0, \beta \geq 0$, L-acceptable iff $\alpha = \beta > 0$.

Frequency fitting is achieved if

$$\alpha^2 = [h^4\mu_0^4\beta^2 - 4h^2\mu_0^2(2-h\mu_0\cot h\mu_0)\beta - 4M]/h^2\mu_0^2M \qquad (15)$$

where $M = M(h\mu_0)$ is given by (14).

The quadratic in β on the right side of (15) factorizes exactly, and it follows that there are real roots for α if $\beta_1 \leq \beta \leq \beta_2$, where

$$\beta_1,\beta_2 = [4 - 2h\mu_0(\cot h\mu_0 \pm \operatorname{cosec} h\mu_0)]/h^2\mu_0^2 .$$

Since $\beta_1 > 0$, frequency fitting with A-acceptability is possible.

There are a number of criteria we could attempt to use in order to choose advantageous non-negative values for the parameters α,β.

(i) <u>Exact damping</u> Assume that $\lambda_0 = 0$, so that the significant frequency is undamped. For $\alpha \geq 0$, $\beta \geq 0$, A-acceptability implies that the frequency-fitted solution will be damped, and this, if severe, may numerically obscure the frequency. An undamped numerical solution is achieved if $|R(ih\mu_0)| = 1$. However, this implies $\beta = 0$ and, by (15), there are no real solutions for α. The best we could do would be to minimize the damping by solving, for given $h\mu_0$, the optimization problem

$$\max_{\alpha,\beta} |R(ih\mu_0)| \quad \text{subject to} \quad \alpha \geq 0,\ \beta \geq 0, \text{ equation (15).}$$

A similar optimization problem can be set up for the case $\lambda_0 \neq 0$.

(ii) <u>Higher order</u> The order of the corresponding method is, in general, 2. Order 3 can be attained if $\beta = \frac{1}{3}$. However, it transpires that with $\beta = \frac{1}{3}$ there are no real solutions of equation (15) for α, for $h\mu_0 < \pi$. Thus we are restricted to order 2.

(iii) <u>L-stability</u> is possible if $\alpha = \beta > 0$. Equation (15) now gives the following quadratic for $\beta(=\alpha)$:

$$h^2\mu_0^2(M-h^2\mu_0^2)\beta^2 + 4h^2\mu_0^2(2-h\mu_0\cot h\mu_0)\beta + 4M = 0$$

It turns out that non-negative real solutions of this quadratic are possible for usable ranges of $h\mu_0$.

<u>References</u>

[1] Ehle, B.L., "On Padé approximations to the exponential function and A-stable methods for the numerical solution of initial value problems", University of Waterloo, Dept. of Applied Analysis and Comouter Science, Research Rep. No. CSRR 2010 (1969).

[2] Cryer, C.W., "A new class of highly stable methods; A_0-stable methods", BIT 13, 153-159, (1973).

[3] Liniger, W. and Willoughby, R.A., "Efficient numerical integration methods for stiff systems of differential equations", IBM Research Report RC-1970 (1967).

FORCED NONLINEAR OSCILLATION FOR CERTAIN THIRD ORDER DIFFERENTIAL EQUATION

By

B. Mehri

1. Let us consider the following non-linear third order differential equation:

$$x''' + f(t, x, x', x'') = e(t) \qquad (1)$$

It is assumed that f is a continuous function of its arguments, and the forcing function e(t) is also a continuous function of t. In this paper, it will be shown that if f(t, x, x', x'') satisfies some conditions given as below, there exist at least one non-trivial solution of (1) which satisfies the following boundary conditions

$$x^{(i)}(0) + x^{(i)}(\omega) = 0 \qquad (2)$$

$$i = 0, 1, 2,$$

In the following, we need a lemma which is known as Wirtinger's inequality 1 .

Lemma 1. Assume x(t) has continuous (n-1)-th derivative and

$$x(t + \omega) + x(t) = 0 ,$$

then

$$\| x^{(i-1)} \|_{\frac{1}{2}} \leq \left(\frac{\omega}{\Pi}\right)^{n-i+1} \cdot \| x^{(x)} \|_{\frac{1}{2}} \qquad (3)$$

$$i = 1, 2, \ldots, n$$

where

$$\| y \|_{1/p} = \left(\int_0^\omega | y(t) |^p dt \right)^{1/p} . \qquad (4)$$

In the three dimensional (x, x', x'') - space, we define Π_3 as a cube satisfying the following conditions

$$\Pi_3 = \{ (x, x', x'') \mid \; |x^{(i-1)}| \leq C/2 \cdot \Pi \cdot \left(\frac{\omega}{\Pi}\right)^{4-i} \; ; i=1,2,3\} \quad (5)$$

where C, is a constant, which we will define it later.

Theorem 1. Assume there exist positive constants k_1, k_2, k_3 and C such that

i) $L = 1 - k_1(\frac{\omega}{\Pi}) - k_2(\frac{\omega}{\Pi})^2 - k_3(\frac{\omega}{\Pi})^3 > 0$

ii) $M \leq L.C$

where

$$M = \operatorname{Max} | F(t, x, x', x'') |$$
$$(x, x', x'') \in \Pi_3 , t \in [0, \omega]$$

and

$$F(t, x, x', x'') = k_1 x'' + k_2 x' + k_3 x + e(t) - f(t, x, x', x'').$$

Then, there exist at least one non-trivial solution of (1) satisfying the boundary conditions (2).

Proof. We consider the following auxiliary boundary value problem

$$x''' + k_1 x'' + k_2 x' + k_3 x = \mu \phi (t, x, x', x'') \qquad (6)$$
$$x^{(i)}(0) + x^{(i)}(\omega) = 0 ; i = 0, 1, 2,$$
$$\mu \in [0, 1]$$

with

$$\phi (t,x,x',x'') = \begin{cases} F(t, x, x', x'') ; \text{ if } (x, x', x'') \in \Pi_3 \\ F(t, \lambda x, \lambda_2 x', \lambda_2 x''); \text{ if } (x, x', x'') \in \Pi_3 \end{cases}$$

where λ_i is such that $\lambda_i x^{(i-1)}$; $i=1, 2, 3$, is on the boundary of the cube Π_3. It can be shown that $\phi(t, x, x', x'')$ is a continuous function of $(x, x',x'') \in R^3$ and furthermore

$$|\phi(t,x,x',x'')| \leq M , 0 \leq t \leq \omega \qquad (7)$$
$$(x, x', x'') \in R^3$$

Let $x(t)$ be any solution of (6). Then from (7) we obtain

$$|x'''| \leq k_1 |x''| + k_2 |x'| + k_3 |x| + \mu . M$$

Now, application of Minkowski's inequality yields

$$||x'''||_{\frac{1}{2}} \leq k_1 || x''||_{\frac{1}{2}} + k_2 || x'||_{\frac{1}{2}} + k_3 || x ||_{\frac{1}{2}} + \mu .M \sqrt{\omega}$$

which by (3), we obtain

$$||x'''||_{\frac{1}{2}} \leq k_1.(\frac{\omega}{\Pi})||x'''||_{\frac{1}{2}} + k_2(\frac{\omega}{\Pi})^2||x'''||_{\frac{1}{2}} + k_3(\frac{\omega}{\Pi})^3||x'''||_{\frac{1}{2}} + \mu.\sqrt{\omega}.M$$

or

$$\{1-k_1(\frac{\omega}{\Pi}) - k_2(\frac{\omega}{\Pi})^2 - k_3(\frac{\omega}{\Pi})^3\} ||x'''||_{\frac{1}{2}} \leq \mu.\ M\sqrt{\omega} \quad .$$

Condition (i) imply that $L = 1-(\frac{\omega}{\Pi})k_1 - (\frac{\omega}{\Pi})^2 k_2 - (\frac{\omega}{\Pi})^3 k_3 > 0$. Hence for $\mu \in [0, 1]$

$$||x'''||_{\frac{1}{2}} \leq \mu \frac{M\sqrt{\omega}}{L} \quad . \tag{8}$$

We now estimate the magnitude of $x^{(i-1)}(t)$, i = 1, 2, 3. We know that

$$x^{(i-1)}(t) = x^{(i-1)}(0) + \int_0^t x^{(i)}(t)dt \ ; \ i=1,\ 2,\ 3$$

and

$$-x^{(i-1)}(0) = \tfrac{1}{2} \int_0^\omega x^{(i)}(t)dt = x^{(i-1)}(\omega); \ i=1,\ 2,\ 3$$

Therefore

$$x^{(i-1)}(t) = \tfrac{1}{2} \int_0^t x^{(i)}(t)dt - \tfrac{1}{2} \int_t^\omega x^{(i)}(t)dt; \ i=1,\ 2,\ 3$$

which by Hölder's inequality, we obtain

$$||x^{(i-1)}||_0 \leq \tfrac{1}{2}\sqrt{\omega} \ . \ ||x^{(i)}||_{\frac{1}{2}} \quad ; \ i=1,\ 2,\ 3$$

or by (3)

$$||x^{(i-1)}||_0 \leq \tfrac{1}{2}\sqrt{\omega}.(\frac{\omega}{\Pi})^{3-i} . ||x'''||_{\frac{1}{2}}$$

here

$$||x^{(i-1)}||_o = \max_{t \in [o,\ \omega]} |x^{(i-1)}(t)| \quad .$$

using inequality (8), we obtain

$$\| x^{(i-1)} \|_o \leq {}^{\mu}/_{2} . {}^{M}/_{2} . \frac{\Pi}{2} \left(\frac{\omega}{\Pi}\right)^{4-i} \quad ; \quad i=1, 2, 3 \tag{9}$$

for $\mu = o$, it follows from (9) that

$$\| x^{(i-1)} \|_o = o \quad ; \; i=1, 2, 3$$

and this means, the problem

$$\begin{aligned} & x''' + k_1 x'' + k_2 x' + k_3 x = o \\ & x^{(i)}(o) + x^{(i)}(\omega) = o \; ; \; i = o, 1, 2 \end{aligned} \tag{10}$$

has only trivial solution, i.e. problem (10) has a Green's function G(t,s). Now, it is obvious that the solution of (6) can be written in the following form

$$x(t) = \mu \int_o^\omega G(t,s) \;\; \phi(s,x(s), x'(s), x''(s)) \, ds . \tag{11}$$

Let s be the space of all continuous functions, with continuous first and second derivatives. Let D be a subset of s such that

$$D = \{ x \in s \;\; | \;\; | x^{(i-1)}(t) | \leq \frac{\Pi}{2} \left(\frac{\omega}{\Pi}\right)^{4-i} . \; C \}$$

$$i = 1, 2, 3$$

and an operator U on D defined as

$$(Ux)(t) = \mu \int_o^\omega G(t, s) \;\; \phi \; (s, x(s), x'(s) \;) \, ds$$

and

$$\frac{d^{i-1}}{dt^{i-1}} (Ux)(t) = \mu \int_o^\omega \frac{\partial^{i-1}}{\partial t^{i-1}} G(t,s) \; \phi \; (s, x(s), x'(s) \, ds$$

$$i = 1, 2, 3$$

we have proved that

$$\left|\frac{d^{i-1}}{dt^{i-1}}(Ux)(t)\right| \leq \frac{\Pi}{2}\cdot\left(\frac{\omega}{\Pi}\right)^{4-i}\cdot\frac{M}{L}$$

$$i = 1, 2, 3$$

Hence, U maps D continuously into itself provided

$$\frac{\Pi}{2}\cdot\left(\frac{\omega}{\Pi}\right)^{4-i}\frac{M}{L} \leq \frac{\Pi}{2}\cdot\left(\frac{\omega}{\Pi}\right)^{4-i}\cdot C$$

i.e.

$$M \leq C.\ L \quad .$$

Therefore, if condition (ii) is satisfied, then U will map D continuously into itself. Hence in the light of Lere - Schauder principle and the complete continuity of the operator defined by (11), this equation has at least one solutioı $\bar{x}(t) \in D$, for $\mu = 1$, we conclude the problem (1) possesses at least one solution $\bar{x}(t)$.

Carrollary 1. With the assumptions of theorem (1), if we further assume that

iii) $f(t, -x, -x', -x'') = -f(t, x, x', x'')$

iiii) $f(t+\omega, x, x', x'') = f(t, x, x', x'')$

V) $e(t+\omega) + e(t) = o$

Then, there exist at least one non-trivial periodic solution of (1) of period 2ω, with mean equal to zero, i.e.

$$\int_{0}^{2\omega} x(t)dt = o$$

Proof. We carry out the 2ω - periodic continuation of $\bar{x}(t)$. Let us define

$$x(t) = \begin{cases} -\bar{x}(t+\omega) & ; \quad -\omega \le t \le 0 \\ \bar{x}(t) & ; \quad 0 \le t \le \omega \\ -\bar{x}(t-\omega) & ; \quad \omega \le t \le 2\omega \end{cases}$$

It is obvious that $x(t)$ is continuous with its first and second derivative and

$$x^{(i)}(t) + x^{(i)}(t+\omega) = 0$$
$$i = 0, 1, 2 \quad .$$

From (iii) and (iiii) we obtain

$$f(t+\omega, x(t+\omega), x'(t+\omega), x''(t+\omega)) = -f(t, x(t), x'(t), x''(t))$$

Hence, if

$$x''' = -f(t, x, x', x'') + e(t) \qquad (12)$$

for all $t \in \left[k\omega, (k+1)\omega\right]$, $k = 0, \pm 1$, (12) remains in force when t is replaced by $t + \omega$. Since (12) is an identity for all $t \in [0,\omega]$, (12) is an identity for all $t \in (-\infty, +\infty)$. We have thus proved the equation

$$x''' + f(t, x, x', x'') = e(t)$$

has a solution $x(t)$ such that $x(t+\omega) = -x(t)$. Hence $x(t)$ is 2ω-periodic and $\int_0^{2\omega} x(t)\,dt = 0$.

2. We assume that all solutions of initial value problems for (1) extend to $[0, T]$.

Under the above assumption we establish the following theorem.

Theorem 2. Let there exist constants $k > 0$ and $C > 0$ such that

$$M \leq k.C \tag{13}$$

where

$$M = \mathrm{Max}\{|kx+e(t) - f(t,x,x',x'')| : t \in [0,T], |x| \leq C, |x'| \leq Ck^{1/3}, |x''| \leq Ck^{2/3}\}.$$

Then there exists ω_o, $0 < \omega_o \leq \frac{2}{k^{1/3}} . \ln\left(\frac{k.C}{M}\right)$ such that for every ω, $0 < \omega \leq \omega_0$ equation (1) has a solution $x(t)$ satisfying the boundary conditions

$$x^{(i)}(0) + x^{(i)}(\omega) = 0, \quad i = 0, 1, 2$$

Proof. Let $\omega \in [0, \omega_0]$, and let $G(t, s)$ be the Green's function

$$G(t,s)=\begin{cases} \frac{1}{(\beta-\alpha)(\gamma-\alpha)} \frac{e^{\alpha(\omega-s+t)}}{1+e^{\alpha\omega}} \frac{1}{(\alpha-\beta)(\gamma-\beta)} \frac{e^{\beta(\omega-s+t)}}{1+e^{\beta\omega}} + \frac{1}{(\beta-\gamma)(\alpha-\gamma)} \frac{e^{\gamma(\omega-s+t)}}{1+e^{\gamma\omega}} & 0 \leq t \leq s \leq \omega \\ \frac{-1}{(\beta-\alpha)(\gamma-\alpha)} \frac{e^{\alpha(t-s)}}{1+e^{\alpha\omega}} \frac{1}{(\alpha-\beta)(\gamma-\beta)} \frac{e^{\beta(t-s)}}{1+e^{\beta\omega}} \frac{1}{(\beta-\gamma)(\alpha-\gamma)} \frac{e^{\gamma(t-s)}}{1+e^{\gamma\omega}} & 0 \leq s \leq t \leq \omega \end{cases}$$

where α, β and γ are the roots of the equation

$$\lambda^3 + k = 0 \tag{14}$$

Then problem (1) with boundary conditions (3) is equivalent to the integral equation

$$x(t) = \int_0^{\omega} G(t,s) \left[kx(s) + e(s) - f(s,x(s), x'(s), x''(s)) \right] ds$$

Let

$$B = \{x(t) \in C''[0,\omega] : |x(t)| \leq 2C, \ |x'(t)| \leq 2Ck^{1/3}, \ |x''(t)| \leq 2Ck^{2/3}\}$$

and define the operator on B by

$$(Ux)(t) = \int_0^{\omega} G(t,s) \left[kx(s) + e(s) - f(s,x(s),x'(s),x''(s)) \right] ds, \quad (15)$$

$$a = \sqrt[3]{k}$$

Then

$$|(Ux)(t)| \leq \frac{2}{k} e^{\frac{a}{2} \cdot \omega} . M$$

and

$$|(Ux)'(t)| \leq \frac{2}{k^{2/3}} e^{\frac{a}{2} \cdot \omega} . M \qquad |(Ux)'(t)| \leq \frac{2}{k^{1/3}} e^{\frac{a}{2} \cdot \omega} . M$$

Hence U maps B continuously into itself provided

$$\omega \leq \frac{2}{a} \ell n \left(\frac{k. C}{M}\right) \quad (16)$$

and, by (13) the right hand side of (16) is non-negative.

Therefore, if $0 < \omega \leq \frac{2}{a} \ell n \left(\frac{k. C}{M}\right)$, it follows from Schauder's theorem that (15) has a solution x(t) such that

$$|x| \leq 2C, \quad |x'| \leq 2C k^{1/3}, \quad |x''| \leq 2C k^{2/3}.$$

Hence (1) has a dolution x(t) satisfying the boundary conditions (3).

Corrallary 2. If in addition to all the hyprotheses of theorem 2, the function f and e satisfy hypòtheses of corrallary 1,then (1) has an 2ω-periodic solution with $\int_0^{2\omega} x(t)dt = 0$.

3. In this section, we shall consider a few applications of our theorems.

A1: Consider the equation

$$x''' + x^3 = \varepsilon \cos t \qquad (17)$$

In this case, we choose $\omega = \Pi$ and $1-k > 0$, with

$$C = \frac{2}{\Pi}\sqrt{\frac{k}{3}}$$

Hence

$$M = \text{Max}\,|kx - x^3 + \varepsilon \cos t\,| = \frac{2}{3}k\sqrt{\frac{k}{3}} + \varepsilon ,$$

and by theorem (1), we must have

$$\left(\frac{2}{3} + \frac{2}{\Pi}\right) k\sqrt{\frac{k}{3}} + \varepsilon \leq \frac{2}{\Pi}\sqrt{\frac{k}{3}} . \qquad (18)$$

Therefore for sufficiently small ε, we can find k such that (18) is satisfied. Hence, we have established the existence of at least one periodic solution of period 2Π with mean equal to zero.

A2: Consider the equation

$$x''' + x^3 + \alpha x'^3 + \beta x''^3 = \varepsilon p(t) \qquad (19)$$

where α, β and ε are positive constants. We want to apply theorem 2 in order to prove (19) has at least one 2ω-periodic solution, provided

$$p(t+\omega) + p(t) = 0 .$$

Let

$$C = \frac{1}{2\sqrt{3}} k^{-\frac{1}{2}}, \text{ then}$$

$$M \leq \frac{1}{3\sqrt{3}} \left[k\sqrt{k} + \frac{\alpha}{\sqrt{k}} + \beta\sqrt{k} \right] + \varepsilon ,$$

and by theorem (2), we must have

$$\frac{1}{3\sqrt{3}} \left[k\sqrt{k} + \frac{\alpha}{\sqrt{k}} + \beta\sqrt{k} \right] + \varepsilon \leq \frac{1}{2\sqrt{3}} \sqrt{k} \qquad (20)$$

It is obvious that for ε sufficiently small that

$$\left(\frac{\beta}{3} - \frac{1}{2}\right)^2 > 4\alpha$$

we can find k such that (20) is satisfied, and hence the results follows from corrollary 2.

REFERENCES

[1] . E.Bechenbach and R. Bellman, "Inequalities," Springer Verlag, Berlin 1962.

[2] . T.Wazewski, "Differential Equations and Their Applications", Praque (1963), pp. 220-242

[3] . B.Mehri, G.G. Hamedani, "On the Existence of Periodic Solutions of Nonlinear Second Order Differential Equations", SIAM. J. Appl. Math. Vol. 29, No.1, 1975.

SUFFICIENT CONDITIONS FOR THE CONVERGENCE, UNIFORMLY IN ε, OF A THREE POINT DIFFERENCE SCHEME FOR A SINGULAR PERTURBATION PROBLEM

John J.H. Miller

Abstract. We consider a general three point difference scheme (P^h) for a two point boundary value singular perturbation problem (P) with a parameter ε, which may be small, and without turning points. We give sufficient conditions for the convergence of the solution of (P^h) to that of (P) as $h \to 0$ with order h, uniformly in ε. We state each step required in the derivation of this result, but we omit the detailed proof of each such step. We remark that, in particular, a well known scheme of Il'in fulfills these conditions and also that several widely used schemes are not convergent uniformly in ε.

Let $\Omega = (a,b)$ and consider the two point boundary problem

$$(P)\qquad \begin{aligned} &Lu \equiv -\varepsilon u'' + a_1 u' + a_0 u = f \quad \text{on } \Omega \\ &u(a),\ u(b) \text{ given} \end{aligned}$$

where the functions a_0, a_1, f are smooth and, for a constant α independent of ε, we have

$$a_0 \geq 0, \quad a_1 \geq \alpha > 0 \quad \text{on } \Omega.$$

The parameter ε is assumed to be positive.

To solve (P) numerically we introduce the uniform mesh $\Omega^h = \{x_i\}_1^{N-1}$, $x_i = a+ih$, $Nh = b-a$, $0 \leq h \leq h_0$ and the three point difference scheme

$$(P^h)\qquad \begin{aligned} &L^h u_i \equiv -a_i u_{i-1} + b_i u_i - c_i u_{i+1} = f_i \qquad 1 \leq i \leq N-1 \\ &u_0,\ u_N \text{ given} \end{aligned}$$

Here $u_i = u^h(x_i)$, $f_i = f^h(x_i)$ are approximations to u and f respectively, and we introduce the notation

$$r_i = (a_i+c_i)/b_i, \quad s_i = (a_i-c_i)/b_i, \qquad \rho = h/\varepsilon.$$

For any mesh function w^h we use the semi-norm

$$||w^h|| = \max_{1 \leq i \leq N-1} |w_i|$$

and we denote the local truncation error in approximating L by the operator L^h by τ^h.

In what follows, unless it is stated to the contrary, all conditions will be assumed to hold for all $1 \leq i \leq N-1$, all $0 < h \leq h_0$ and all $\varepsilon > 0$. In addition, all constants will be assumed to be independent of i, h and ε, and the same symbol may be used to denote different constants.

The following two lemmas are easy to prove.

Lemma 1. (P^h) is consistent with (P) if the following two conditions are fulfilled

(i) $|b_i(1-r_i)-a_0(x_i)|+|hb_is_i-a_1(x_i)|+|h\rho b_ir_i/2-1|+h^3|b_i(r_i\pm s_i)| = o(1)$ as $h \to 0$

(ii) $|f_i-f(x_i)|+|u_0-u(a)|+|u_N-u(b)| = o(1)$ as $h \to 0$.

Lemma 2. L^h is positive iff the following two conditions are fulfilled

(iii) $b_i > 0$

(iv) $|s_i| \leq r_i \leq 1$ with at least one inequality strict.

Recalling that the maximum principle holds for L^h if it is positive, we can prove the following result.

Lemma 3. Assume that, in addition to (iii) and (iv), the following condition holds

(v) $hb_is_i \geq \alpha$.

Then L^h is stable with stability constant $1/\alpha$ and we have the following estimates for the solutions u, u_i of (P), (P^h) respectively

(1) $|u_i| \leq \max\{|u_0|,|u_N|\} + \frac{1}{\alpha}\,||f^h||$

(2) $|u_i-u(x_i)| \leq \max\{|u_0-u(a)|,|u_N-u(b)|\} + \frac{1}{\alpha}(||f^h-f||+||\tau^hu||)$.

The classical convergence result is contained in

Theorem 1. Conditions (i)-(v) imply that the solution of (P^h) converges to that of (P) as $h \to 0$ for each fixed ε.

For convergence uniformly in ε we need a further estimate for u_i.

Lemma 4. Assume that, in addition to (iii)-(v), for some constant β, we have

(vi) $b_i(1-r_i) \leq \beta$

(vii) $s_i \geq r_i \tanh \alpha\rho/2$.

Then

$$|u_{i+2}-u_{i-2}| \leq C_1C_2(h+e^{-\alpha(1-x_i-2h)/\varepsilon})$$

where

$$C_1 = \max\{|u_0|,|u_N|\} + \frac{1}{\alpha}\,||f^h||, \quad C_2 = \max\{2,1+\beta/\alpha\}.$$

Following Il'in [1], we use the next two lemmas as the basis of our proof of uniform convergence. The first is a general principle.

Lemma 5. Let F^h be any quantity depending on h, p a positive number and C a constant independent of h. Then

$$|F^h-F| \le C\,h^p \quad \text{for} \quad 0 < h \le h_0$$

iff the following two conditions hold

(1) $|F^h-F| = o(1)$ as $h \to 0$

(2) $|F^h-F^{h/2}| \le C\,h^p$ for $0 < h \le h_0$.

The second is a well known technique involving the maximum principle.

Lemma 6. Assume that, for any difference operator L^h and any mesh function w_i, the following two conditions hold

(1) the maximum principle holds for L^h

(2) there exists a (comparison) function ϕ_i such that $|L^h w_i| \le L^h \phi_i$, $|w_0| \le \phi_0$, $|w_N| \le \phi_N$.

Then

$$|w_i| \le \phi_i.$$

Using the last two lemmas we obtain immediately the following convergence result

Theorem 2. Let u, u_i be the solutions of (P), (P^h) respectively.
Assume that there is a constant C_0, a positive number p and a comparison function ϕ_i for which the following five conditions hold

(1) u^h converges to u as $h \to 0$ for each fixed ε.

(2) the maximum principle holds for L^h.

(3) $|(u^h-u^{h/2})(a)| \le \phi_0$, $|(u^h-u^{h/2})(b)| \le \phi_N$.

(4) $|L^h(u^h-u^{h/2})(x_i)| \le L^h\phi_i$.

(5) $|\phi_i| \le C_0 h^p$.

Then

$$|u_i-u(x_i)| \le C_0 h^p.$$

We remark at once that this theorem gives sufficient conditions for convergence with order h^p, uniformly in ε, provided that the constant C_0 in (5) is independent of ε and h.

In what follows we determine sufficient conditions on the parameters of (P^h) for conditions (1)-(5) of Theorem 2 to be fulfilled with $p = 1$.

We see immediately from Lemma 2 and Theorem 1 that (1) and (2) are satisfied if conditions (i)-(v) hold. Sufficient conditions for (3)-(5) to hold are established by means of the following ten lemmas.

We begin by introducing the mesh function

$$\psi_i = h(1+x_i+e^{-\alpha(1-x_i-h)/2\varepsilon})$$

concerning which we have

Lemma 7. Assume that, in addition to (iii) and (iv), for some constant γ, we have

(viii) $s_i \leq \gamma \tanh \alpha\rho/2$.

Then the following hold

(1) $\psi_i \leq 3h$

(2) $\psi_0 \geq h$, $\psi_N \geq h$

(3) $L^h\psi_i \geq C_3 hb_i s_i(h+s_i e^{-\alpha(1-x_i-h)/2\varepsilon})$

where

$$C_3 = \min\{1, 1/2\gamma^2\}.$$

We assume for the moment that there is a constant C_4 such that the following two conditions hold

(A) $\quad |(u^h-u^{h/2})(a)| \leq C_4 h, \quad |(u^h-u^{h/2})(b)| \leq C_4 h$

(B) $\quad |L^h(u^h-u^{h/2})(x_i)| \leq C_4 hb_i s_i(h+s_i e^{-\alpha(1-x_i-h)/2\varepsilon})$.

We then define the comparison function

$$\phi_i = (C_4/C_3)\psi_i$$

and, using (A), (B) and Lemma 7, we see that conditions (3)-(5) of Theorem 2 are fulfilled with $p = 1$ and $C_0 = 3C_4/C_3$. It remains therefore to determine sufficient conditions for (A) and (B) to hold.

(A) certainly holds if, for example, there is a constant δ such that

(ix) $\quad |(u^h-u^{h/2})(a)| \leq \delta|u(a)|h, \quad |(u^h-u^{h/2})(b)| \leq \delta|u(b)|h$.

To find sufficient conditions for (B) to hold we consider, for convenience only, $L^{2h}(u^{2h}-u^h)(x_i)$ for $x_i \in \Omega^{2h}$, and we show that (B) holds with h replaced by 2h. For quantities defined on the coarse mesh Ω^{2h} we use a superscript 2h, while the lack of such a superscript means that the quantity is associated with Ω^h. We introduce the notation

$$A_i = a_i/b_i, \quad C_i = c_i/b_i, \quad F_i = f_i/b_i.$$

The symbols C', C" will indicate constants which are not necessarily the same at each occurrence. D_+, D_- denote respectively the forward and backward divided differences and $D_0 = (D_+ + D_-)/2$.

Using the equation in (P^h) we can prove

Lemma 8. We have

$$M_i L^{2h}(u^{2h}-u^h)(x_i)/b_i^{2h} = (M_i F_i^{2h}-2F_i)+(F_i-A_i F_{i-1}-C_i F_{i+1})+(M_i C_i^{2h}-C_i C_{i+1})u_{i+2}+$$
$$+(M_i A_i^{2h}-A_i A_{i-1})u_{i-2}$$

where

$$M_i = 1-A_iC_{i-1}-A_{i+1}C_i.$$

Additional algebraic manipulations give

Lemma 9. We have

$$M_iL^{2h}(u^{2h}-u^h)(x_i)/b_i^{2h} = M_i(f_i^{2h}-f_i)/b_i^{2h}+(M_i/b_i^{2h}-2/b_i)f_i+$$
$$+[(1-r_i)/2b_{i+1}+h(s_i-1/2)D_+(1/b_i)]f_{i+1}+[(1-r_i)/2b_{i-1}+h(s_i+1/2)D_-(1/b_i)]f_{i-1}+$$
$$+2hs_iD_0f_i/b_i-h^2D_+D_-f_i/2b_i+X_i(u_{i+2}+u_{i-2})/2+Y_i(u_{i+2}-u_{i-2})/2.$$

where

$$X_i = 1-r_i^2-(1-r_i^{2h})M_i+hs_iD_0r_i-h^2r_iD_+D_-r_i/2$$

$$Y_i = s_i-s_i^{2h}M_i-(1-r_i)s_i-hr_iD_0r_i/2-hs_iD_0s_i/2+h^2s_iD_+D_-r_i/4+h^2r_iD_+D_-r_i/4.$$

Lemma 10. If (iv) holds we have

$$1/4 \le M_i \le 1.$$

Using Lemmas 3,4,9 and 10 we can prove

Lemma 11. Assume that, in addition to (iii)-(vii), for some constant C', we have

(x) $|f_i^{2h}-f_i| \le C'h\,||f^h||$

(xi) $|f_i|+|D_0f_i|+|D_+D_-f_i| \le C'\,||f^h||.$

Then there is a constant C" such that

$$|L^{2h}(u^{2h}-u^h)(x_i)| \le C''C_1b_i^{2h}[h^2(s_i+s_i^{2h})+|M_i/b_i^{2h}-2/b_i|+h|D_+(1/b_i)|+h|D_-(1/b_i)|$$
$$+|X_i|+|Y_i|(h+e^{-\alpha(1-x_i-2h)/\varepsilon})].$$

Lemma 12. We have

$$2M_i = (1+s_i^2)+(1-r_i^2)-h^2r_iD_+D_-r_i/2-hr_iD_0s_i+hs_iD_0r_i+h^2s_iD_+D_-s_i/2.$$

From Lemma 12 we obtain immediately

Lemma 13. Assume that, in addition to (iv)-(vi), for some constant C', we have

(xii) $|D_0r_i|+h|D_+D_-r_i| \le C'hs_i$

(xiii) $|D_0s_i|+|D_+D_-s_i| \le C's_i.$

Then there is a constant C" such that

$$|2M_i-(1+s_i^2)| \le C''hs_i.$$

Using Lemma 13 we then have

Lemma 14. Assume that, in addition to (iv)-(vi), (xii) and (xiii), for some constant C', we have

(xiv) $|(1+s_i^2)s_i^{2h}-2s_i| \le C'hs_i^{2h}(h+s_i^{2h})$

(xv) $|(1+s_i^2)(1-r_i^{2h})-2(1-r_i^2)| \le C'h^2 s_i^{2h}$

(xvi) $|(1+s_i^2)/b_i^{2h}-4/b_i| \le C'h^2 s_i^{2h}$.

Then there is a constant C" such that

(1) $s_i \le C''s_i^{2h}$

(2) $|M_i s_i^{2h}-s_i| \le C''hs_i^{2h}(h+s_i^{2h})$

(3) $|M_i(1-r_i^{2h})-(1-r_i^2)| \le C''h^2 s_i^{2h}$

(4) $|M_i/b_i^{2h}-2/b_i| \le C''h^2 s_i^{2h}$.

From Lemmas 9 and 14 we then get

Lemma 15. If (iv)-(vi) and (xii)-(xvi) hold, then for some constant C"

$$|X_i| \le C''h^2, \quad |Y_i| \le C''hs_i^{2h}(h+s_i^{2h}).$$

Finally, combining Lemmas 11, 14 and 15 we have

Lemma 16. Assume that, in addition to (iii)-(vii) and (x)-(xvi), for some constant C', we have

(xvii) $|D_+(1/b_i)|+|D_-(1/b_i)| \le C'hs_i$

then (B) holds.

Summarizing, we have shown that the solution of (P^h) converges to the solution of (P) as $h \to 0$, uniformly in ε and with order h, if conditions (i)-(xvii) are fulfilled.

It is then a straightforward matter to check, for example, that Il'in's difference scheme

$$-\varepsilon(a_1(x_i)\rho/2)\coth(a_1(x_i)\rho/2)D_+D_-u_i+a_1(x_i)D_0u_i+a_0(x_i)u_i = f(x_i)$$

$$u_0 = u(a), \quad u_N = u(b)$$

fulfills these condition. We have thus generalized the convergence result given in [1], where it is assumed that $a_0(x) \equiv 0$. Moreover, it is not hard to construct families of difference schemes which satisfy conditions (i)-(xvii).

We remark finally that a necessary condition for convergence, uniformly in ε, for a class of difference schemes, is established in [2]. There it is shown that several common difference schemes are not convergent uniformly in ε.

REFERENCES

[1] A.M. Il'in, "Differencing scheme for a differential equation with a small parameter affecting the highest derivative", Math. Notes Acad.Sci. USSR 6 (1969), 596-602.
[2] J.J.H. Miller, "Some finite difference schemes for a singular perturbatio problem" in Constructive Function Theory. Proc.Int.Conf. Constructive Function Theory, Blagoevgrad 30 May - 4 June 1977, Sofia (in print).

EXPERIENCES ON NUMERICAL CALCULATION OF FIELDS

W. Müller

1. Significance of field calculations

Numerical solutions of Maxwell's equations are of utmost importance for all applications in energy-technique. Since the properties of the electric and magnetic fields are described exhaustingly and exactly by the field equations and the characteristics of all materials used are well known, the field quantities within the electric machines under investigation can be calculated with high accuracy. This was possible because of the progress of numerical mathematics in the past 2o years and it can be utilized, as fast computers are available today.

The knowledge of field distribution is the basis for all further evaluations, therefore the possibility is open to calculate the properties of a new machine before it is realized. This is an advantage which cannot be overestimated. Indeed the comparison of measured and calculated fields has shown a very good agreement in all cases investigated. [1]

In the performance of field calculations some difficulties arise which are due to special features of the problems to be solved. This is to be pointed out here for magnetostatic fields, it is however true for electrostatic and eddy-current field problems.

2. Way of solution

The basic equations of the magnetostatic field are the following:

$$(1) \quad \text{curl}\,\vec{H} = \vec{J}$$

$$(2) \quad \text{div}\,\vec{B} = 0$$

$$(3) \quad \vec{B} = \mu\cdot\vec{H}, \quad \mu = f_i(|\vec{H}|)$$

$\vec{B}$ is the magnetic flux density, $\vec{H}$ the magnetic field strength and the permeability μ is a given function of the unknown field strength $\vec{H}$, which differs from material to material. The distribution of the

magnetizable material is also given. The current density $\vec{J}$ can be easily found from the spacious arrangement of the conductors and the currents belonging to them. Furthermore, on the edge of the domain in which the field is to be calculated, boundary conditions are prescribed. The nonlinear field problem defined by (1) to (3) can be solved by a number of methods. In the following only one method is taken into account which is characterized by the fact that μ is recalculated from time to time using the "old" approximation of the field and the characteristics of the materials. Therefore μ can be regarded as a function of space and the field problem is linear.
To solve the linearized equations (1) to (3) $\vec{B}$ is usually derived from a vector potential $\vec{A}$. (see e.g. [2],[3])

(4) $\vec{B} = \text{curl}\, \vec{A}$

This leads to the following differential equation for the system variable $\vec{A}$:

(5) $\text{curl}\, \frac{1}{\mu}\, \text{curl}\, \vec{A} = \vec{J}$

Unfortunately this equation can be solved easily only for two-dimensional geometries. In this case two of the three components of $\vec{A}$ are zero and for the remaining component a system of linear difference equations can be derived which fulfills all sufficient conditions for application of the SOR iteration method.
In three dimensions all three components of $\vec{A}$ are nonzero and the system of difference equations has such a complicated structure that it cannot be solved with reasonable computer time. Therefore it is more advantageous to use the method of scalar potential given in [4]. The field $\vec{H}$ is split up in two parts

(6) $\vec{H} = \vec{H}_i + \vec{H}_p$

The first part, $\vec{H}_i$, has to fulfill the rotational equation (1), but no condition to the sources is imposed to $\vec{H}_i$.

(7) $\text{curl}\, \vec{H}_i = \vec{J}$, $\text{div}\, \vec{H}_i$ = arbitrary function

This partitioning is not unique, but for every choice of $\vec{H}_i$ the curl of the second part, $\vec{H}_p$, is zero and $\vec{H}_p$ can be deduced from a scalar potential Φ.

(8) $\vec{H}_p = -\,\text{grad}\, \Phi$

One way to calculate a suitable $\vec{H}_i$ is given by the formula following from Green's theorem:

$$(9) \quad \vec{H}_i(\vec{r}) = \frac{1}{4\pi} \int_{V'} \frac{\vec{J}(\vec{r}\,') \times (\vec{r}-\vec{r}\,')}{|\vec{r}-\vec{r}\,'|^3} dV'$$

The cost for tabulating (9) at the mesh points is rising with the square of the number of points N and exceeds soon the capacity even of very fast computers. A better way is to calculate $\vec{H}_i$ numerically as described in [5]. The scalar potential Φ fulfills the elliptic, self adjoint differential equation

$$(1o) \quad \text{div } \mu \text{grad } \Phi = \text{div } \mu \vec{H}_i$$

which can be solved much easier than the corresponding equation (5) for the vector potential.

3. Discretisation

For the calculation of magnetic fields in the interior of rotating machines the finite difference method has proved to be well suited, as both the materials and the conductors are arranged in such a way that it can be easily described by the coordinate planes of the mesh. In the domain V a mesh is laid generated by three sets of coordinate planes not neccessarily equidistant. Both cartesian and cylindrical coordinates occur. The derivation of difference equations has been described many times in literature, therefore it is only outlined here for further understanding.

The equations (5) resp. (1o) to be solved are integrated over a region V' which encloses the mesh point P_o under consideration. (see fig. 1) In the case of vector potential Stoke's theorem and in the case of scalar potential Gauss's theorem is applied. The result for the vector potential is:

$$(11) \quad \int_{V'} \text{curl } \frac{1}{\mu} \text{ curl } \vec{A} \cdot d\vec{A} = \int_{V'} \vec{J} \cdot d\vec{A} = I$$

$$(12) \quad \int_{V'} \frac{1}{\mu} \text{ curl } \vec{A}\, d\vec{s} = I$$

and for the scalar potential:

$$(13) \quad \int_{V'} \text{div } \mu \text{grad } \Phi dV = \int_{V'} \text{div } \mu \vec{H}_i \, dV$$

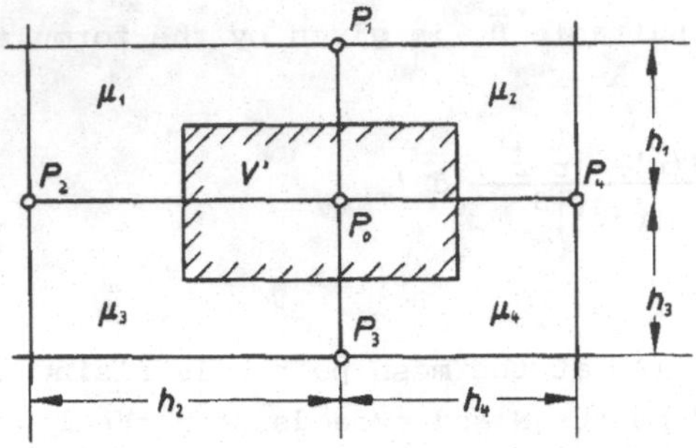

Fig. 1a: Derivation of difference equation, 2-dim. case

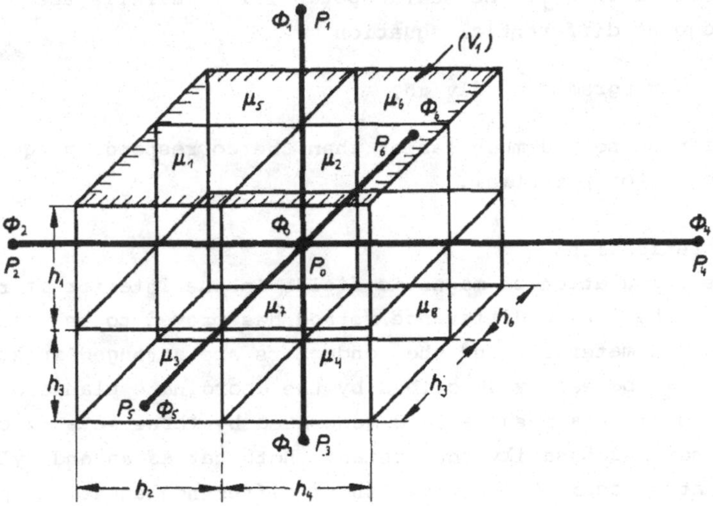

Fig. 1b: Derivation of difference equation, 3-dim. case

$$(14) \quad \int_{(V')} \frac{\partial \Phi}{\partial n}\, dA = \int_{(V')} H_{in}\, dA$$

n is the direction of the normal vector pointing outward and (V') the boundary of V'. To evaluate the line- resp. surface integrals, following simplifications are usual:

a) The permeability μ is regarded to be constant in the interior of elementary mesh regions.
b) The derivatives $\partial\Phi/\partial n$, $\partial A/\partial n$ for example at the part (V_1) of the surface of fig. 1b are replaced by the difference quotients $(\Phi_1 - \Phi_0)/h_1$, $(A_1 - A_0)/h_1$.

This leads finally to a linear equation between the unknowns Φ_0, A_0

at the point P_o and Φ_i, A_i at the neighbor points P_i, which are now denoted by x_i.

$$(15) \quad \alpha_o x_o = \sum_1^6 \alpha_i x_i + c$$

The coefficients α_i satisfy the relations

$$(16) \quad \alpha_o = \sum_1^6 \alpha_i , \quad \alpha_i > o$$

4. Iterative solution of the linear system

Numbering the mesh points by lines and writing down the difference equations in this order, a linear system is obtained.

$$(17) \quad A\vec{x} = \vec{c}$$

$\vec{x} = (x_i)$ is the vector of the unknowns, $\vec{c} = (c_i)$ the vector of the right sides. The N×N matrix $A = (a_{ij})$ is a symmetric and diagonal-dominant L-matrix and possesses Youngs property A. [6] The iterative solution of (17) can be carried out by using the SOR-method.

$$(18) \quad x_i^{(n+1)} = x_i^{(n)} + \omega \left(\sum_{j=1}^{i-1} b_{ij} x_j^{(n+1)} + \sum_{j=i+1}^{N} b_{ij} x_j^{(n)} + c_i - x_i^{(n)} \right)$$

which is equivalent to the matrix notation

$$(19) \quad \vec{x}^{(n+1)} = L_\omega \vec{x}^{(n)} + \vec{c}$$

$B = (b_{ij})$ is the Jacobi-matrix associated to A, L_ω the SOR iteration matrix and n the index of iteration. The optimal iteration parameter ω_b is given by:

$$(2o) \quad \omega_b = \frac{2}{1+\sqrt{1-\rho^2(B)}}$$

$\rho(B)$ being the spectral radius of B. The spectral radius of L_ω defined by (19) is given by:

$$(21) \quad \rho(L_\omega) = \omega_b - 1$$

All this is well known. In practical cases N is about some 1o ooo and $\rho(B)$ very close to unity. The range

$$.999 < \rho(B) < 1$$

is typical and corresponds to the range

$$.92 < \rho(L_\omega) < 1$$

The improvement of convergence and the reduction of computing time are considerable.

6. The condition of periodicity

Some features of field problems not yet considered here make it necessary to modify the way of solution. Firstly the condition of periodicity in the φ-direction may exist. Then an additional connection between the first and the last point of each line appears and the structure of the matrix A is changed as shown by the graph of fig. 2a.

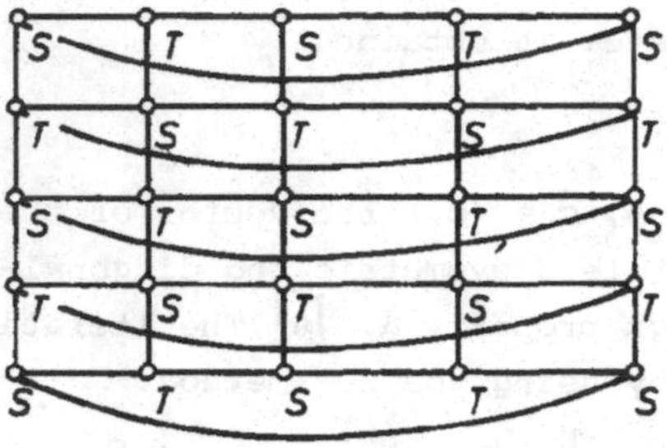

Fig 2a: Graph of the matrix A. Property A does not exist

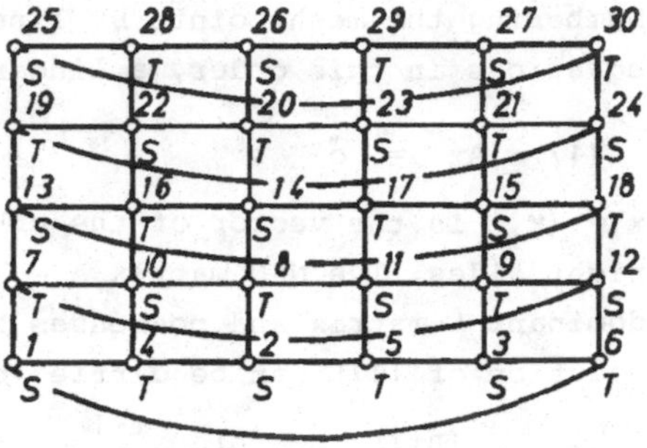

Fig. 2b: Inserting an additional mesh column to get the property A

When the number of the columns of fig. 2a is odd, then A has not the property A. This can be easily overcome by inserting an additional mesh column (fig. 2b), but the linewise numbering of the unknowns is not consistent. A consistent order of the mesh points is shown in fig. 2b, a generally applicable way is given in (7). The increase of convergence, due to reordering of mesh points, is shown in fig. 3,

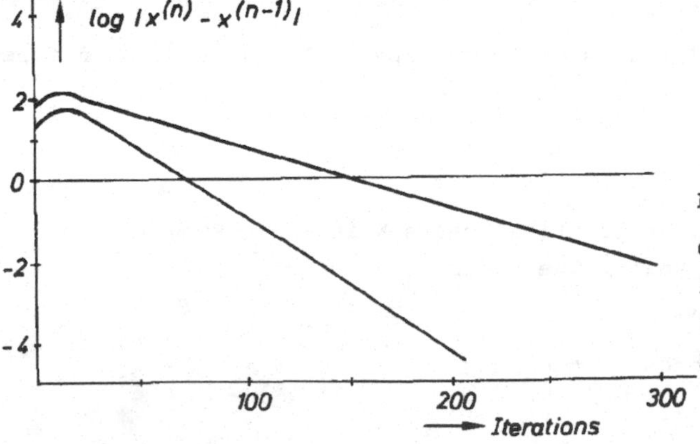

Fig. 3:

Convergence of SOR-iteration.

1: consistent order

2: no consistent order

which is representative for many examples. The convergence of the SOR-method applied to a non consistent matrix is better than the convergence of Gauss-Seidel- iteration, however, the convergence with consistent order is not attained.

6. Modification of the SOR iteration procedure

A second complication is caused by an extension of the mesh which changes also the structure of the coefficient matrix A. For the calculation of field in the interior of rotating machines it is indispensable to use a mesh in which the radial coordinate is subdivited into a number of intervals as shown in fig. 4.

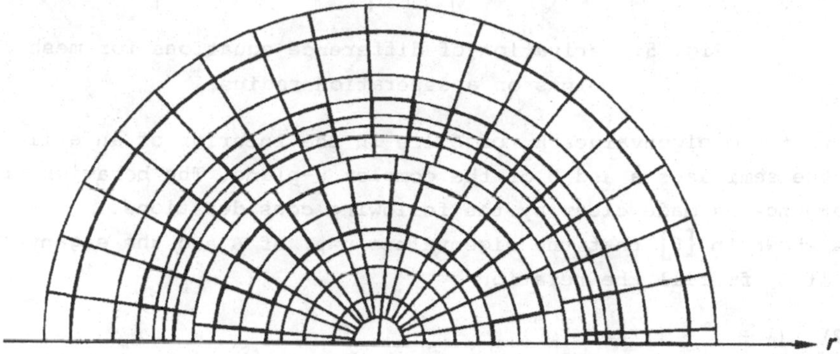

Fig. 4: Mesh with one seperation radius

In each subregion the lines φ = const can be chosen independently. By this means the number of mesh points is reduced considerably. The difference equations for points on the "seperation" radii cannot derived in the way wich is used in a regular mesh. One of the vertical neighbor point needed for the difference quotient does not exist in the mesh. The potential at the missing point must be interpolated from the potentials at the two adjacent points P_1', P_1'' resp. P_3', P_3'' of fig. 5. This results in a 6-point difference equation instead of the usual 5-point formula. In three dimensions a 8-point formula is obtained instead of the 7-point formula. By this extension the matrix A has lost the property A definitely and therefore the Block-SOR iteration must be used.

Furthermore A is not symmetric and so the eigenvalues of the Block-Jacobi matrix B associated to A are generally complex. This is not presumed in the derivation of (2o). The results of the extension of Young's theory to nonsymmetric matrices A are given in [7] for the

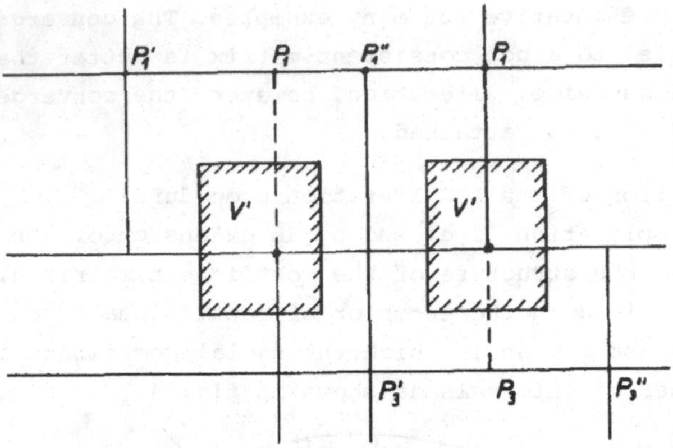

Fig. 5: Derivation of difference equations for mesh points on a seperation radius

case that the eigenvalues μ_i of B lie in the interior of an ellipse with the semi-axes a and b of the complex μ-plane. The behavior of convergence is made clear by the following considerations.
It is shown in [6] that the eigenvalues $\mu=\mu_i$ of B and the eigenvalues $\lambda=\lambda_i$ of L_ω fulfill the relation

$$(22) \quad (\lambda + \omega - 1)^2 = \mu^2\omega^2\lambda$$

Setting $\sigma=\sqrt{\lambda}$ and taking the square root on both sides gives:

$$(23) \quad \sigma^2 + \omega - 1 = \mu\,\omega\,\sigma$$

The circle $\sigma = re^{i\varphi}$ (i=imag. unit) of the σ-plane is transformed in the ellipse

$$(24) \quad \begin{aligned} \mathrm{Re}(\mu) &= \frac{1}{\omega}\left(r + \frac{\omega-1}{r}\right)\cos\varphi = a\cos\varphi \\ \mathrm{Im}(\mu) &= \frac{1}{\omega}\left(r - \frac{\omega-1}{r}\right)\sin\varphi = b\sin\varphi \end{aligned}$$

of the μ-plane. This ellipse is also the image of the circle with radius $r' = (\omega-1)/r$. Therefore the greater of the two values r_1, r_2 given by

$$(25) \quad \begin{aligned} r_1 &= \frac{1}{2}\omega a + \sqrt{\frac{1}{4}\omega^2 a^2 - (\omega-1)} \\ r_2 &= \frac{1}{2}\omega b + \sqrt{\frac{1}{4}\omega^2 b^2 + (\omega-1)} \end{aligned}$$

is the spectral radius of L_ω. This is shown in fig. 6. The convergence rate can be taken from this figure by combining the two curves which

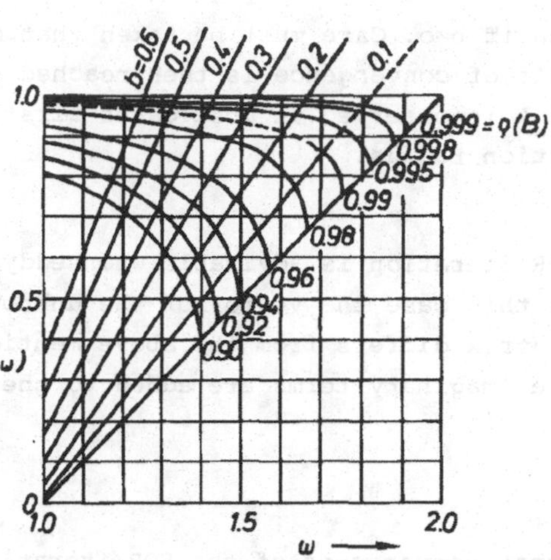

Fig. 6

The behavior of the asymtotic convergence rate of the SOR iteration applied to a linear system with a nonsymmetric coefficient matrix A

belong to the given values of a and b. For example the convergence rate for the set a = .99, b = .1 is marked in fig. 6 by a dotted line. It is seen that the iteration can diverge in the interval $1<\omega<2$. The evaluation of a,b can be done in the following way. Fig. 6 shows that the SOR iteration always converges for $\omega=1$ and if $b>o$ always diverges for $\omega=2$. The convergence rate $\rho(L_1)$ and the divergence rate $\rho(L_2)$ are found by performing some trial iterations. From these a and b can be calculated according to

$$(26) \quad a = \sqrt{\rho(L_1)}$$

$$(27) \quad b = \frac{1}{2\sqrt{\rho(L_2)}}(\rho(L_2) - 1)$$

Sometimes it is better to set $\omega=\omega_o$ and compute $\rho(L_{\omega_o})$ instead of $\rho(L_2)$. Then b is given by:

$$(28) \quad b = \frac{\rho(L_{\omega_o}) - (\omega_o-1)}{\omega_o\sqrt{\rho(L_{\omega_o})}}$$

The best ω is taken from

$$(29) \quad \omega_b = \frac{2}{1+\sqrt{1-a^2+b^2}}$$

and the spectral radius from

$$(3o) \quad \rho(L_{\omega_o}) = (\omega_b-1)\frac{a+b}{a-b}$$

The symmetry of the matrix A is not a necessary condition for applying SOR, but the gain of convergence in comparison to the Gauss-Seidel iteration is lower if b>o than if b=o. Care must be taken that ω is not overestimated because the limit of convergence is then reached soon. In practice the unsymmetry of A is observed in most cases firstly, if there is more than one seperation radius.

7. Iteration with complex ω.

A further extension of the SOR iteration is advisable when eddy-current problems are to be solved. In this case the vector of the unknowns is complex and the coefficient matrix differs from the above mentioned one only by the fact that pure imaginary terms are added to the entries of the main diagonal.

$$(31) \quad a'_{jj} = a_{jj} + i\beta_j$$

It has been observed in [9] tbat convergence of the SOR iteration can be considerably improved if a complex iteration parameter ω is used, which was estimated from trial and experience. In [1o] it was found that best convergence is achieved when the following value for ω is used:

$$(32) \quad \omega_b = \frac{2}{1+\sqrt{1-\mu_{max}^2}}$$

μ_{max} is the absolut greatest complex eigenvalue of the Jacobi-matrix associated to A. Because μ_{max} is complex a complex value of ω is obtained from (32). The similarity of the equations (32) and (2o) makes it probable that (32) can be proved rigorously.

References

1 W. Fritz: Anwendung neuer numerischer Methoden bei der Berechnung von Magnetfeldern, Wirbelstromsystemen und Temperaturfeldern in elektrischen Maschinen. VDE-Fachtagung, Nov. 1976, Mannheim

2 E.A. Erdelyi, E.F. Fuchs: Nonlinear magnetic field analysis of dc-machines. Part 1-3, Inst. Electr. and Electronics Eng., Trans. on power Appl. and Systems, Vol. 89 (197o) p. 1546-1583.

3 K. Reichert: Über ein numerisches Verfahren zur Berechnung von Magnetfeldern und von Wirbelströmen in elektrischen Maschinen. Arch. Elektrotechn. Vol. 52 (1968/69) p. 176-194.

4 W. Müller, W. Wolff: Numerische Berechnung dreidimensionaler Magnetfelder für grosse Turbogeneratoren bei feldabhängiger Permeabilität und beliebiger Stromdichteverteilung. ETZ Vol. 94 (1973) p.276-282.

5 W. Müller, W. Wolff: Beitrag zur numerischen Berechnung von Magnetfeldern. ETZ Vol. 96 (1975) p. 269-273.

6 D. M. Young: Iterative methods for solving partial difference equations of elliptic type. Trans. Amer. Soc.,Vol.76(1954) p. 92-111.

7 J. Krüger: Vergleichende Untersuchung von einigen iterativen Verfahren..... ,Diplomarbeit, TH-Darmstadt, Fachbereich 4.

8 E.L. Wachspress: Iterative Solution of Elliptic Systems....,
Prentice Hall Inc. (1964).

9 S.A. Nasar, L. del Cid: Propulsion and levitation forces in a single sided linear induction motor..., Proc. IEEE, Vol. 61,No.5 (May 1973).

1o W. Müller: Berechnung von Feldern und Kräften in Linearmotoren. Wiss. Ber. AEG-Telefunken, 49 (1976) 3, p. 1o8-116.

An Application of the Differential Equations of the Sound Ray

By K. Nixdorff

Summary: Sound ranging of detonations or similar short sounds in the atmosphere may be done in a mathematically very simple manner if only sound rays travelling along the surface of the ground are considered. The evaluation of sound rays which are raised temporarily to higher layers of the atmosphere is rather complicated. The following exposition describes the difficulties and the attempts to deal with them.

1. Rectilinear Sound Rays

Ray acoustics is applicable in sound ranging in the atmosphere (in the following text, the appendix "in the atmosphere" will be omitted for brevity). In this sense the terms sound ray, wave front etc. are used. Some other terms will be explained immediately (see Figure 1): A basis consists of two microphones (e.g. M_1 and M_2) and a system is composed of several bases. A short sound originating from the sound source S reaches the microphone M_i after travelling the time t_i. The differences

$$t_{ij} := t_i - t_j \,, \qquad i \neq j$$

are measured. Let us assume that all microphones and the sound source have the same height and that all sound rays travelling from the sound source to the microphones are rectilinear and travel with the constant sound speed c. Now let us choose an orthonormal-rectilinear coordinate system such that the microphone M_i has the coordinates (-b/2;0), the microphone M_j the coordinates (b/2;0), and the sound source S the coordinates $(x_s;y_s)$. Simple geometrical considerations show that the sound source S is lying on the hyperbola

$$\left(\frac{2x_s}{c\, t_{ij}}\right)^2 - \left(\frac{2y_s}{\sqrt{b^2-c^2t_{ij}^2}}\right)^2 = 1 \,. \tag{A}$$

The sign of the difference t_{ij} determines the branch of the hyperbola on which the sound source S is lying. In practice the sign of the ordinate y_s is known, so that only half of one branch of the hyperbola has to be taken as a locus for the position of the sound source S.

The loci derived in that way from several bases will usually not intersect at one point, which would be the position of the sound source S, due to errors in the measurements of the coordinates, the time differences, and the sound velocity. Each pair of loci will intersect at one point and the distribution of these points may produce a fair impression of the approximate location of the sound source S.

The influence of temperature and humidity is considered. Let us denote by c_o the velocity of sound in dry air at a temperature of 0° Celsius and at an air pressure of 101 325 Pascal, by c the velocity of sound at a temperature ϑ and at

Figure 1

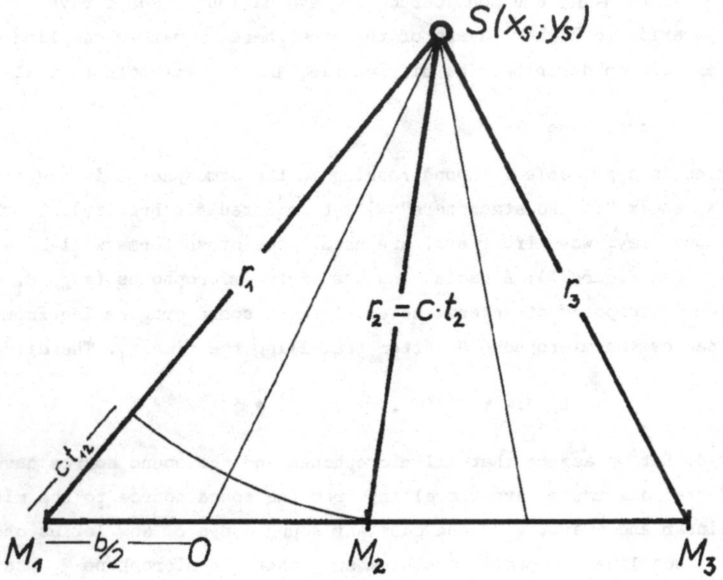

a ratio e/p of the partial water vapour pressure in air to the total air pressure, α the coefficient of expansion of ideal gases referred to one degree Celsius, M_1 the average molecular weight of dry air, and M_2 the molecular weight of water. Then the velocity c is given according to [1] by

$$c = c_o\left(1 + \frac{\alpha}{2}\vartheta + \frac{1}{2}\,\frac{M_1 - M_2}{M_1}\,\frac{e}{p}\right)$$

or

$$c = \left(331.3 + 0.6079\,\vartheta + 62.98\,\frac{e}{p}\right)\ \text{m/s}$$

with

ϑ in degrees Celsius, e and p same units.

At 0° Celsius, 101 325 Pascal, and at a relative humidity of 75 %, e/p is approximately 0.00452.

The influence of a constant wind velocity $\vec{w} = (w_x, w_y)$ leads to

$$\left(\frac{2x_2}{ct_{ij} - b\frac{w_x}{c}}\right)^2 - \left(\frac{2y_s}{\sqrt{b^2 - \left(ct_{ij} - b\frac{w_x}{c}\right)^2}}\right)^2 = 1 \qquad \text{(B)}$$

for

$$\frac{|\vec{w}|}{c} \ll 1$$

which is practically always satisfied.

2. Derivation of the Differential Equations of the Bent Sound Rays

If the sound rays do not travel rectilinearly the equations (A) and (B) are no longer immediately applicable, and it is necessary to use the differential equations of the bent sound rays.

These equations may be derived using analytical mechanics, as shown by L.D. Landau and E.M. Lifschitz [2]. E. Esclangon [3] and R. Sänger [4] gave elementary derivations using Huygens' principle. The following derivation is based on that of E. Esclangon, but is more rigorous and shorter than the preceding three. (All symbols are used in accordance with Figure 2.)

In the presence of wind the sound ray is no longer orthogonal to the wave front. Let us denote by $\vec{n}$ a unit vector orthogonal to the wave front and pointing into the direction of the propagation of the sound wave, and by $\vec{r}$ the radius vector of the general point of the sound ray. We consider the transportation of a sufficiently small element of the wave front during the time Δt. This element is assumed to be plane, perpendicular to the drawing plane, and containing the points P and Q. The

Figure 2

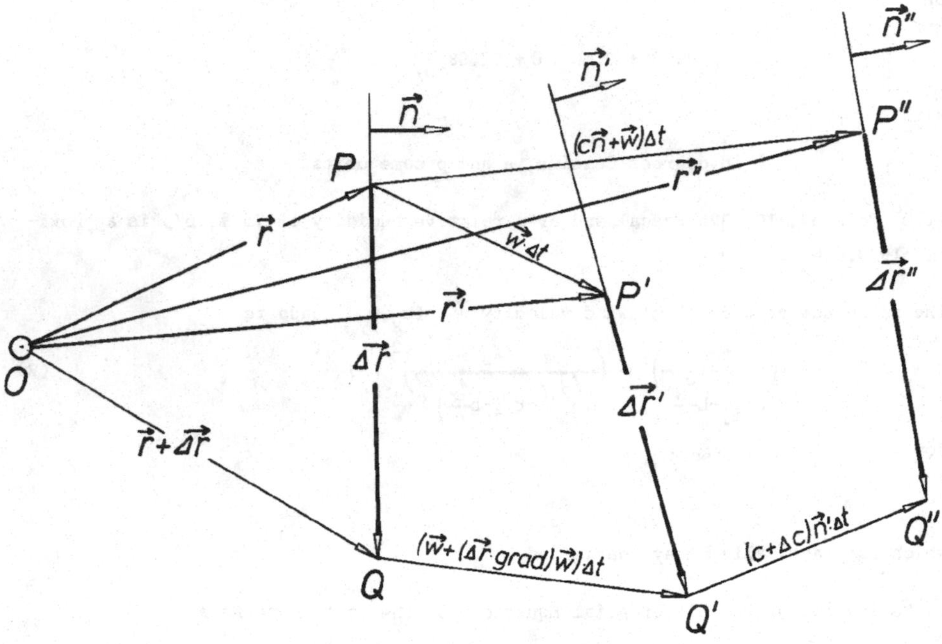
$\vec{n}''$
$\vec{n}'$
$\vec{n}$
P''
$(c\vec{n}+\vec{w})\Delta t$
P
$\vec{r}''$
$\vec{r}$
$\vec{w}\cdot\Delta t$
P'
$\Delta\vec{r}''$
$\vec{r}'$
O
$\Delta\vec{r}$
$\Delta\vec{r}'$
$\vec{r}+\Delta\vec{r}$
Q''
$(c+\Delta c)\vec{n}'\Delta t$
$(\vec{w}+(\Delta\vec{r}\cdot grad)\vec{w})\Delta t$
Q
Q'

wind velocity and the sound velocity c transport the point P to the point P", and the point Q to the point Q". The following equations may be taken from Figure 2:

$$\Delta\vec{r}\,\vec{n} = 0, \tag{1}$$

$$\vec{r}\,' + \Delta\vec{r}\,' = \vec{r} + \Delta\vec{r} + [\vec{w} + (\Delta\vec{r}\ \mathrm{grad})\vec{w}]\Delta t, \tag{2}$$

$$\Delta\vec{r}\,'\,\vec{n}\,' = 0, \tag{3}$$

$$\vec{n}\,' = \vec{n} + \dot{\vec{n}}_W\,\Delta t, \tag{4}$$

$$\Delta\vec{r}\,''\,\vec{n}\,'' = 0, \tag{5}$$

$$\vec{n}\,'' = \vec{n} + \dot{\vec{n}}\,\Delta t, \tag{6}$$

$$\vec{r}\,'' = \vec{r} + (c\,\vec{n}+\vec{w})\Delta t, \tag{7}$$

$$\vec{r}\,'' + \Delta\vec{r}\,'' = \vec{r}\,' + \Delta\vec{r}\,' + (c+\Delta c)\vec{n}\,'\,\Delta t. \tag{8}$$

The equations (4) to (8) give

$$[\vec{r}\,' + \Delta\vec{r}\,' + (c+\Delta c)(\vec{n}+\dot{\vec{n}}_W\,\Delta t)\Delta t - \vec{r} - (c\vec{n}+\vec{w})\Delta t](\vec{n}+\dot{\vec{n}}\,\Delta t) = 0 \tag{9}$$

which, with equation (2) yields

$$\{\Delta r + [(\Delta\vec{r}\ \mathrm{grad})\vec{w} + (c+\Delta c)(\vec{n}+\dot{\vec{n}}_W\,\Delta t) - c\vec{n}]\Delta t\}(\vec{n}+\dot{\vec{n}}\,\Delta t) = 0\ . \tag{10}$$

From the equations (10) and (1) one obtains after division by Δt

$$\Delta\vec{r}\,\dot{\vec{n}} + [(\Delta\vec{r}\ \mathrm{grad})\vec{w} + (c+\Delta c)(\vec{n}+\dot{\vec{n}}_W\,\Delta t) - c\,\vec{n}](\vec{n}+\dot{\vec{n}}\,\Delta t) = 0. \tag{11}$$

This equation reduces for $\Delta t \to 0$ to

$$\Delta\vec{r}\,\dot{\vec{n}} + \vec{n}\,[(\Delta\vec{r}\ \mathrm{grad})\vec{w}] + \Delta c = 0. \tag{12}$$

If

$$\Delta c = \Delta\vec{r}\ \mathrm{grad}\ c \tag{13}$$

is introduced into (12) and the vector identity

$$\vec{a}[(\vec{b}\ \mathrm{grad})\vec{c}] = \vec{b}[(\vec{a}\ \mathrm{grad})\vec{c} + \vec{a}\times\mathrm{curl}\ \vec{c}]$$

is used, one finds

$$\Delta\vec{r}[\dot{\vec{n}} + (\vec{n}\ \mathrm{grad})\vec{w} + \vec{n}\times\mathrm{curl}\ \vec{w} + \mathrm{grad}\ c] = 0. \tag{14}$$

Comparison of the equations (1) and (14) shows that

$$\dot{\vec{n}} + (\vec{n}\ \mathrm{grad})\vec{w} + \vec{n}\times\mathrm{curl}\ \vec{w} + \mathrm{grad}\ c = C\,\vec{n}, \tag{15}$$

where C is obtained by scalar multiplication of equation (15) by $\vec{n}$ to be

$$C = \vec{n}[(\vec{n}\ \mathrm{grad})\vec{w} + \mathrm{grad}\ c]. \tag{16}$$

The change of the sound ray is described according to equation (7) by

$$\dot{\vec{r}} = c\,\vec{n} + \vec{w}. \tag{17}$$

The equations (15) to (17) are the differential equations of the bent sound ray.

3. Stratified Atmosphere

Now let

$$\vec{r} = (x,y,z)$$
$$\vec{n} = (\cos\tau \cos\varphi,\ \cos\tau \sin\varphi,\ \sin\tau).$$

In view of the obtainable weather data let

$$\vec{w} = (u(z),\ v(z),\ 0),$$
$$c = c(z).$$

The equations (15) to (17) become

$$\dot{\Phi} \cos^2 \tau = 0, \tag{18}$$

$$\dot{\tau} \cos\tau = -\cos^2\tau\left[\cos\tau\left(\frac{du}{dz}\cos\varphi + \frac{dv}{dz}\sin\varphi\right) + \frac{dc}{dz}\right], \tag{19}$$

$$\dot{x} = c\cos\tau\cos\varphi + u, \tag{20}$$

$$\dot{y} = c\cos\tau\sin\varphi + v, \tag{21}$$

$$\dot{z} = c\sin\tau. \tag{22}$$

If the uninteresting case $\tau = \pi/2$ is disregarded the equations (18) and (19) yield

$$\varphi = \mathrm{const.} \tag{23}$$

$$\frac{c}{\cos\tau} + u\cos\varphi + v\sin\varphi = \mathrm{const.} \tag{24}$$

The equation (24) determines the relation between φ and z. The equations (23) and (24) inserted into the equations (20) to (22) result in

$$x = x_0 + \int_{z_0}^{z} \frac{c\cos\tau\cos\varphi + u}{c\sin\tau}\,dz \tag{25}$$

$$y = y_0 + \int_{z_0}^{z} \frac{c\cos\tau\sin\varphi + v}{c\sin\tau}\,dz, \tag{26}$$

$$t = t_0 + \int_{z_0}^{z} \frac{dz}{c\sin\tau}. \tag{27}$$

Under the above assumptions the maximum height of a bent sound ray is characterized by

$$\tau = 0.$$

Therefore the integrals in the equations (25) to (27) become improper ones if integrated up to or including the maximum height.

Let us think about using the equations (23) and (24) and either (20) to (22) or (25) to (27) to determine the coordinates of the sound source S.

Since these coordinates are unknown the sound rays have to be calculated starting from the microphones. At each microphone M_i there will be needed as initial values the coordinates (x_i,y_i,z_i), the direction of the normal given by the angles (φ_i,τ_i), and the time t_i needed by the sound to travel from the microphone M_i to the sound source S or vice versa. The angles φ_i and τ_i are not measurable with sufficient accuracy because of the microstructure of the wave front. The time t_i is unknown, too. Only the differences t_{ij} are known.

If the system consists of four microphones, then 24 values $[(x_i,y_i,z_i,\varphi_i,\tau_i,t_i), i = 1(1)4]$ would be needed for straightforward integration. The twelve coordinates of the four microphones are known. The three time differences measured by the three bases are also known. It is further known that the unknown sound rays go to the same point, the location of the sound source S. This gives nine conditions for the coordinates of the sound rays coming from the microphones M_i at the different times t_i. These are 24 conditions. Thus the coordinates of S might be determined, but this has not been proven yet.

If all microphones and the sound source are located at the same height z, then proceeding as above might show that three microphones are sufficient. This has also not been proven yet.

4. Constant Gradients

In practical sound ranging a rule of thumb is needed to evaluate without the aid of a computer the signals of the bent sound rays. There are several of such rules. The following one is distinguished by its strict deduction and by its simple result.

The exposition follows the one given by E. Esclangon [3], improved by R. Sänger [4], and adds some further improvements.

Consider one sound ray, starting at the time t = 0 from

$$x = y = z = 0,\ \varphi = 0,\ \tau = \tau_o\ , \tag{28}$$

and arriving at the time $t = T$ at a point with

$$x = X, \; y = Y, \; z = Z, \; \varphi = \varphi_e, \; \tau = \tau_e \; . \tag{29}$$

In sound ranging the sound rays considered will usually stay below 600 m. Therefore one can set

$$c = c_B + a_c z, \tag{30}$$

$$u = (w_B + a_w z) \cos \omega_g \tag{31}$$

$$v = (w_B + a_w z) \sin \omega_g \tag{32}$$

with

$$c_B, \; a_c, \; w_B, \; a_w, \; \omega_g \quad \text{all constant,}$$

and ω_g chosen in such a way that a_w is not negative. These assumptions, especially ω_g = const., are not always sufficiently satisfied. In this case the results obtained may be too inaccurate.

Usually the inequalities

$$|a_c| < 0.01 \; s^{-1}$$
$$a_w < 0.03 \; s^{-1}$$

hold. The wind may turn with increasing height, in general to the right on the northern hemisphere. In Central Europe a change of the direction (measured in radians) of up to 0.1 for every 100 m difference of height will be normal.

Inserting the equations (30) to (32) into the equations (19) to (22) results in a system of differential equations which can be solved in closed form. With

$$w = w_B + a_w z$$

the differential equations obtained are

$$\frac{dt}{d\tau} = - \frac{1}{(a_c + a_w \cos \omega_g \cos \tau) \cos \tau} , \tag{33}$$

$$\frac{dx}{d\tau} = - \frac{c \cos \tau + w \cos \omega_g}{(a_c + a_w \cos \omega_g \cos \tau) \cos \tau} , \tag{34}$$

$$\frac{dy}{d\tau} = - \frac{w \sin \omega_g}{(a_c + a_w \cos \omega_g \cos \tau) \cos \tau} , \tag{35}$$

$$\frac{dz}{d\tau} = - \frac{c \sin \tau}{(a_c + a_w \cos \omega_g \cos \tau) \cos \tau} . \tag{36}$$

The integration of the differential equations (33) to (36) with the initial conditions (28) may use the result of integrating the differential equation (36) to integrate the differential equations (34) and (35). The solutions are

$$t = \frac{1}{a_c}\left[\operatorname{arth}\sin\tau_o - \operatorname{arth}\sin\tau + a_w\cos\omega_g\int_{\tau_o}^{\tau}\frac{d\tau}{a_c+a_w\cos\omega_g\cos\tau}\right], \quad (37)$$

$$x = x_1 + x_2\cos\omega_g, \quad (38)$$

$$y = x_2\sin\omega_g, \quad (39)$$

$$z = \frac{c_B(\cos\tau-\cos\tau_o)}{(a_c+a_w\cos\omega_g)\cos\tau_o}, \quad (40)$$

with

$$x_1 = -\frac{c_B(a_c+a_w\cos\omega_g\cos\tau_o)}{(a_c^2-a_w^2\cos^2\omega_g)\cos\tau_o}\left\{a_c\left[\frac{\sin\tau}{a_c+a_w\cos\omega_g\cos\tau}-\frac{\sin\tau_o}{a_c+a_w\cos\omega_g\cos\tau_o}\right]\right.$$

$$\left.-\,a_w\cos\omega_g\int_{\tau_o}^{\tau}\frac{d\tau}{a_c+a_w\cos\omega_g\cos\tau_o}\right\},$$

$$x_2 = \frac{c_B a_w - w_B a_c}{a_c^2}(\operatorname{arth}\sin\tau - \operatorname{arth}\sin\tau_o) +$$

$$+\frac{c_B a_w^2\cos\omega_g(a_c+a_w\cos\omega_g\cos\tau_o)}{a_c\left(a_c^2-a_w^2\cos^2\omega_g\right)\cos\tau_o}\left[\frac{\sin\tau}{a_c+a_w\cos\omega_g\cos\tau}-\frac{\sin\tau_o}{a_c+a_w\cos\omega_g\cos\tau_o}\right]$$

$$+\,a_w\frac{-c_B a_c^3+[w_B a_c(a_c^2-a_w^2\cos^2\omega_g)-c_B a_w(2a_c^2-a_w^2\cos^2\omega_g)]\cos\omega_g\cos\tau_o}{a_c^2\left(a_c^2-a_w^2\cos^2\omega_g\right)\cos\tau_o}\cdot A$$

with

$$A = \int_{\tau_o}^{\tau}\frac{d\tau}{a_c+a_w\cos\omega_g\cos\tau}.$$

The remaining integral in the above expressions can be evaluated in closed form. One obtains

$$\int_{\tau_o}^{\tau}\frac{d\tau}{a_c+a_w\cos\omega_g\cos\tau} = \frac{2}{\sqrt{a_c^2-a_w^2\cos^2\omega_g}}\arctan\frac{\sqrt{a_c^2-a_w^2\cos^2\omega_g}\left(\tan\frac{\tau}{2}-\tan\frac{\tau_o}{2}\right)}{a_c+a_w\cos\omega_g+(a_c-a_w\cos\omega_g)\tan\frac{\tau}{2}\tan\frac{\tau_o}{2}}$$

for $a_c > a_w\cos\omega_g$,

$$\int_{\tau_o}^{\tau}\frac{d\tau}{a_c+a_w\cos\omega_g\cos\tau} = \frac{1}{a_c}\left(\tan\frac{\tau}{2}-\tan\frac{\tau_o}{2}\right)$$

for $a_c = a_w \cos \omega_g$, and

$$\int_{\tau_o}^{\tau} \frac{d\tau}{a_c+a_w \cos \omega_g \cos \tau} = \frac{2}{\sqrt{a_w^2 \cos^2 \omega_g - a_c^2}} \operatorname{arth} \frac{\sqrt{a_w^2 \cos^2 \omega_g - a_c^2}\left(\tan \frac{\tau}{2} - \tan \frac{\tau_o}{2}\right)}{a_c+a_w \cos \omega_g - \left(a_w \cos \omega_g - a_c\right)\tan\frac{\tau}{2} \tan\frac{\tau_o}{2}}$$

for $a_c < a_w \cos \omega_g$.

The maximum height H of the sound range is characterized by the elevation $\tau = 0$. From equation (4) follows

$$H = \frac{c_B(1-\cos \tau_o)}{(a_c+a_w \cos \omega_g)\cos \tau_o} . \tag{41}$$

According to the equations (29) and (4)) the sound ray reaches the ground level $z = 0$ again at

$$\tau = \tau_e = - \tau_o , \tag{42}$$

(42) inserted into the equations (37) to (39) determines the quantities T,X and Y:

$$T = \frac{1}{a_c}\left[2 \operatorname{arth} \sin \tau_o + a_w \cos \omega_g \int_{\tau_o}^{-\tau_o} \frac{d\tau}{a_c+a_w \cos \omega_g \cos}\right] , \tag{43}$$

$$X = X_1 + X_2 \cos \omega_g , \tag{44}$$

$$Y = X_2 \sin \omega_g , \tag{45}$$

with

$$X_1 = \frac{c_B}{a_c^2-a_w^2 \cos^2 \omega_g}\left[2 a_c \tan \tau_o + \frac{a_w \cos \omega_g (a_c+a_w \cos \omega_g \cos \tau_o)}{\cos \tau_o} \int_{\tau_o}^{-\tau_o} \frac{d\tau}{a_c+a_w \cos \omega_g \cos \tau}\right] .$$

$$X_2 = - 2 \frac{c_B a_w - w_B a_c}{a_c^2} \operatorname{arth} \sin \tau_o - \frac{2 c_B a_w^2 \cos \omega_g}{a_c\left(a_c^2-a_w^2 \cos \omega_g\right)} \tan \tau_o$$

$$+ a_w \frac{-c_B a_c^3 + \left[w_B a_c\left(a_c^2-a_w^2\cos^2\omega_g\right) - c_B a_w\left(2a_c^2-a_w^2\cos^2\omega_g\right)\right]\cos \omega_g \cos \tau_o}{a_c^2\left(a_c^2-a_w^2\cos^2 \omega_g\right)\cos \tau_o} \int_{\tau_o}^{-\tau_o} \frac{d\tau}{a_c+a_w \cos \omega_g \cos \tau} .$$

The remaining integral in the above expressions can be evaluated in closed form. One obtains

$$\int_{\tau_o}^{-\tau_o} \frac{d\tau}{a_c+a_w \cos \omega_g \cos \tau} = - \frac{2}{\sqrt{a_c^2-a_w^2\cos^2\omega_g}} \arctan \frac{2\sqrt{a_c^2-a_w^2\cos^2 \omega_g} \tan \frac{\tau_o}{2}}{a_c+a_w \cos \omega_g - (a_c-a_w \cos \omega_g)\tan^2 \frac{\tau_o}{2}}$$

for $a_c > a_w \cos \omega_g$,

$$\int_{\tau_o}^{-\tau_o} \frac{d\tau}{a_c+a_w \cos \omega_g \cos \tau} = - \frac{2}{a_c} \tan \frac{\tau_o}{2}$$

for $a_c = a_w \cos\omega_g$, and

$$\int_{\tau_o}^{\tau_o} \frac{d\tau}{a_c+a_w \cos\omega_g \cos} = -\frac{2}{\sqrt{a_w^2\cos^2\omega_g-a_c^2}}\,\mathrm{arth}\,\frac{2\sqrt{a_w^2\cos^2\omega_g-a_c^2}\,\tan\frac{\tau_o}{2}}{a_c+a_w\cos\omega_g+(a_w\cos\omega_g-a_c)\tan^2\frac{\tau_o}{2}}$$

for $a_c < a_w \cos\omega_g$.

In sound ranging the elevation τ_o is usually smaller than 0.4 (measured in radians), so the right sides of the equations (41) and (43) to (45) are replaced by Taylor series about the value $\tau_o = 0$. The series are truncated in such a way that the resulting polynomials are sufficiently accurate for

$$T < 100\ \mathrm{s},$$

which is satisfactory for sound ranging. The formulas obtained are

$$H = \frac{c_B\tau_o^2}{2(a_c+a_w\cos_g)}\left(1+\frac{5}{12}\tau_o^2\right) \tag{46}$$

$$T = \frac{2\tau_o}{a_c+a_w\cos\omega_g}\left[1+\frac{a_c+2a_w\cos\omega_g}{a_c+a_w\cos\omega_g}\,\frac{\tau_o^2}{6} + \frac{5a_c^2+15a_ca_w\cos\omega_g+16a_w^2\cos^2\omega_g}{(a_c+a_w\cos\omega_g)^2}\,\frac{\tau_o^4}{120}\right] \tag{47}$$

$$X_1 = \frac{2c_B\tau_o}{a_c+a_w\cos\omega_g}\left[1+\frac{2a_c+a_w\cos\omega_g}{a_c+a_w\cos\omega_g}\,\frac{\tau_o^2}{6} + \frac{16a_c^2+23a_ca_w\cos\omega_g+5a_w^2\cos^2\omega_g}{(a_c+a_w\cos\omega_g)^2}\,\frac{\tau_o^4}{120}\right] \tag{48}$$

$$X_2 = \frac{2\tau_o}{a_c+a_w\cos\omega_g}\left[w_B+\frac{2c_Ba_w+w_B(a_c+2a_w\cos\omega_g)}{a_c+a_w\cos\omega_g}\,\frac{\tau_o^2}{6} + \frac{4c_Ba_w(5a_c+7a_w\cos\omega_g)+w_B(5a_c^2+15a_ca_w\cos\omega_g+16a_w^2\cos^2\omega_g)}{(a_c+a_w\cos\omega_g)^2}\right]\frac{\tau_o^4}{120} \tag{49}$$

The time T, but not the elevation τ_o , can be measured with sufficient accuracy. Thus the quantities H, X and Y are needed as functions of the time T. To obtain these functions, the function (47) is inverted to get the elevation τ_o as a function of the time T and the result is inserted into the functions (46), (48) and (49). This yields

$$\tau_o = \frac{1}{2}(a_c+a_w \cos\omega_g)T\left[1-(a_c+2a_w \cos\omega_g)(a_c+a_w \cos\omega_g)\frac{T^2}{24}\right.$$

$$\left.+\left(5a_c^2+25a_c a_w \cos\omega_g+24a_w^2 \cos^2\omega_g\right)(a_c+a_w \cos\omega_g)^2 \frac{T^4}{1920}\right], \tag{50}$$

$$H = \frac{c_B}{8}(a_c+a_w \cos\omega_g)T^2\left[1+(a_c-3a_w \cos\omega_g)(a_c+a_w \cos\omega_g)\frac{T^2}{48}\right], \tag{51}$$

$$X_1 = c_B T\left[1+\left(a_c^2-a_w^2 \cos^2\omega_g\right)\frac{T^2}{24}\right.$$

$$\left.+\left(a_c^2-2a_c a_w \cos\omega_g+9a_w^2 \cos^2\omega_g\right)(a_c+a_w \cos\omega_g)^2 \frac{T^4}{1920}\right], \tag{52}$$

$$X_2 = T\left[w_B+c_B a_w(a_c+a_w \cos\omega_g)\frac{T^2}{12} - c_B a_w^2 \cos\omega_g(a_c+a_w \cos\omega_g)^2 \frac{T^4}{160}\right]. \tag{53}$$

The equations (52) and (54) together with the equations (44) and (45) give

$$X = c_B T\left[1 + \frac{w_B \cos\omega_g}{c_B} + (a_c+a_w \cos\omega_g)^2 \frac{T^2}{24}\right.$$

$$\left.+ (a_c-3a_w \cos\omega_g)(a_c+a_w \cos\omega_g)^3 \frac{T^4}{1920}\right]. \tag{54}$$

$$Y = T \sin\omega_g\left[w_B+c_B a_w(a_c+a_w \cos\omega_g)\frac{T^2}{12} - c_B a_w^2 \cos\omega_g(a_c+a_w \cos\omega_g)^2\frac{T^4}{160}\right] \tag{55}$$

In the following, the time T will be restricted to

$$T < 60\ \mathrm{s}$$

which allows to neglect the last term in each of the expressions (51) to (55).

This leads to

$$X^2 + Y^2 = \left\{c_B^2+w_B^2+2c_B w_B \cos\omega_g+2\left[(c_B+w_B \cos\omega_g)(a_c+a_w \cos\omega_g)\right.\right.$$

$$\left.\left.+ 2w_B a_w \sin^2\omega_g\right]\frac{H}{3}\right\}T^2 . \tag{56}$$

The equation (56) is equivalent to

$$X^2 + Y^2 = \left[\left(c_B+a_c \frac{H}{3}\right)^2 + \left(w_B+a_w \frac{H}{3}\right)^2 + 2\left(c_B+a_c \frac{H}{3}\right)\left(w_B+a_w \frac{H}{3}\right)\cos\omega_g\right]T^2 =$$

$$= \left[c^2+w^2+2c\, w \cos\omega_g\right]\Big|_{z=H/3} T^2 \tag{57}$$

if terms proportional to one of the following products:

$$a_c^2 ,\ a_w^2 ,\ a_c\, a_w ,\ w_B\, a_w ,$$

are neglected.

Under the assumption that any difference in height between sound source and microphone is negligible the equation (57) gives the rectilinear distance between sound source and microphone. Equation (57) leads to equation (B) under the assumption that c and $\vec{w}$ are taken at the "effective sound weather height" H/3.

The equations (51) and (57) are jointly the promised rule of thumb, termed "averaging over the height", which must be used iteratively, starting with the weather data at the ground. It may be necessary to switch to the equations (51), (54) and (55) after a few steps to improve the accuracy.

There are other rules of thumb. One is called "gradient correction" and is equivalent to the above one, as shown by R. Sänger [4]. Another one is the "stratified heights procedure". This assumes that the effective sound weather height will be obtained by averaging over all layers of the atmosphere with the length of the sound ray in each layer as its weight factor. Usually, this results in an effective sound weather height of 2H/3 or more. So this rule is plausible, but incorrect.

5. Fermat's Principle

There are a few attempts to use a computer in sound ranging. They all refer to Fermat's principle. This principle, as stated e.g. by L.D. Landau and E.M. Lifschitz [2] is easy to derive. We start with the original formulation for a sound ray travelling from the radius rector $\vec{r}_1$ to the radius vector $\vec{r}_2$:

$$\int_{\vec{r}_1}^{\vec{r}_2} dt = \text{Min} \tag{58}$$

and use the ray equation (17), written as

$$dt = \frac{c\,\vec{n}-\vec{w}}{c^2-\vec{w}^2}\,d\vec{r}\;. \tag{59}$$

The ray equation (17) yields also

$$c\,\sin(\vec{n},\dot{\vec{r}}) = |\vec{w}|\sin(\vec{w},\dot{\vec{r}}), \tag{60}$$

which after multiplication by $d\vec{r}$ leads to

$$\sqrt{c^2\,d\vec{r}^2 - (c\,\vec{n}\,d\vec{r})^2} = \sqrt{\vec{w}^2\,d\vec{r}^2 - (\vec{w}\,d\vec{r})^2}, \tag{61}$$

or

$$c\,\vec{n}\,d\vec{r} = \sqrt{(c^2-\vec{w}^2)d\vec{r}^2 + (\vec{w}\,d\vec{r})^2}\;, \tag{62}$$

which together with the equation (59) transforms the equation (58) into

$$\int_{\vec{r}_1}^{\vec{r}_2} \frac{\sqrt{(c^2-\vec{w}^2)d\vec{r}^2+(\vec{w}\,d\vec{r})^2}-\vec{w}\,d\vec{r}}{c^2-\vec{w}^2} = \text{Min.} \tag{63}$$

Using the equation (17) transforms the variational principle (63) so that the integration occurs with respect to the time t. Considering the obtainable weather data this seems to be of no advantage.

For a stratified atmosphere the variational principle (63) becomes $\left(\frac{dx}{dz} = x', \frac{dy}{dz} = y'\right)$:

$$\int_{z_1}^{z_2} \frac{\sqrt{c^2(x'^2+y'^2+1)-(uy'-vx')^2-u^2-v^2}-ux'-vy'}{c^2-u^2-v^2}\,dz = \text{Min}\,. \tag{64}$$

This might be used with integration from one end point of the sound ray to its maximum height.

The equation (24) determines the height z as a function of τ. Therefore the equations (23) and (24) allow to render the variational principle (63) to $\left(\frac{dx}{d\tau} = x', \frac{dy}{d\tau} = y', \frac{dz}{d\tau} = z'\right)$:

$$\int_{\tau_1}^{\tau_2} \frac{\sqrt{c^2(x'^2+y'^2+z'^2)-(uy'-vx')^2-(u^2+v^2)z'^2}-ux'-vy'}{c^2-u^2-v^2}\,d\tau = \text{Min}, \tag{65}$$

which allows integration over the whole sound ray.

Up to now, Fermat's principle has not been used in sound ranging without further assumptions to simplify calculations. It has not been shown yet that in sound ranging Fermat's principle may be more useful than using the differential equations of the bent sound ray directly.

Literature: [1] K. Nixdorff,
Mathematische Methoden der Schallortung in der Atmosphäre,
Braunschweig 1977;

[2] L.D. Landau and E.M. Lifschitz,
Lehrbuch der theoretischen Physik, Bd. VI: Hydrodynamik,
Berlin 1966;

[3] E. Esclangon,
L'acoustique des canons et des projectiles,
Mémorial de l'Artillerie Française, Paris 1925;

[4] R. Sänger,
Artilleristische Schallmessung,
Zürich 1938.

On using the Du Fort Frankel scheme for determination of the velocity profile in turbulent boundary layer along an oscillating wall

Syvert P.Nørsett

1. Introduction

In order to study the behavior of a ship in open sea and its effects on the ship itself, a very simple model is set up.

An infinite circular cylinder of radius r is placed in a fluid of viscousity ν and density ρ. (See Figure 1.) The cylinder is oscillating with a tangential velocity of $v_0\cos(\omega t)$, $t \geq o$. Just above the surface, in the boundary layer, the viscosity predominates the behavior of the fluid and hence the velocity profile. The viscous effect in the proximity of the surface may be estimated as follows.

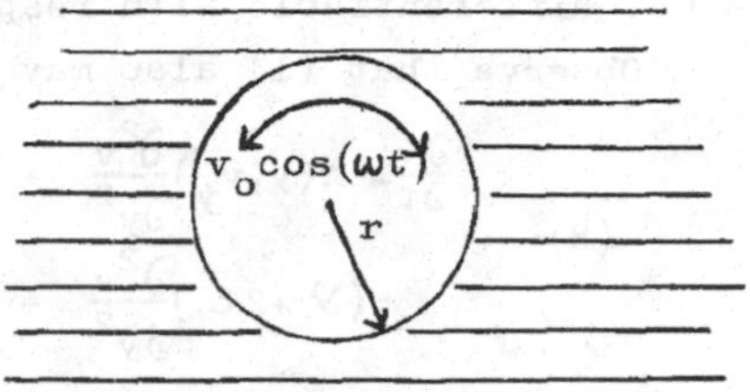

Figure 1.

Consider an infinite flat plate undergoing simple harmonic oscillation parallel to the plate (See Figure 2.). It can now be shown that the mean local velocity $v(y,t)$ is determined from the parabolic equation

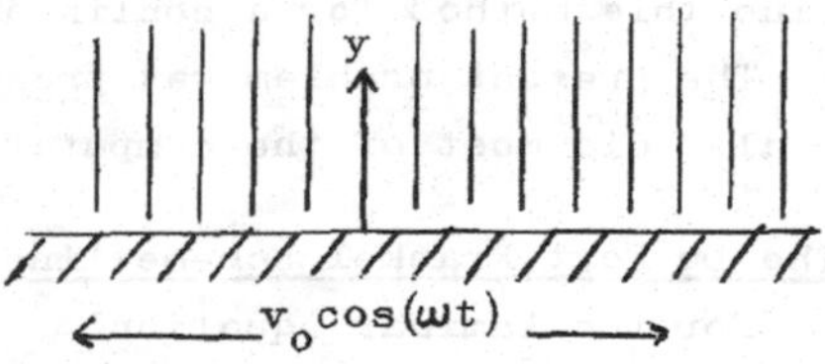

Figure 2.

$$(1) \qquad \frac{\partial v}{\partial t} = \frac{\partial}{\partial y}\left((\nu + \varepsilon)\frac{\partial v}{\partial y}\right), \quad y > 0, \; t > 0,$$

where

$$(2) \qquad \varepsilon = \kappa^2 y^2 \exp(-2y/A)\left|\frac{\partial v}{\partial y}\right|$$

$$(3) \qquad A = 26\sqrt{\nu} \Big/ \sqrt{\left|\frac{\partial v}{\partial y}(0,t)\right|},$$

(κ is called the Van Karman constant.) and the boundary- and initia conditions

$$(4) \qquad v(0,t) = v_0\cos(\omega t), \quad t \geq 0$$

$$(5) \qquad \lim_{y\to\infty} v(y,t) = 0$$

$$(6) \qquad v(y,0) = v_0\exp(-By/v_0), \quad \text{B a given constant.}$$

In the laminar case, when $\varepsilon=0$, the problem is called Stokes second problem. For $\varepsilon\neq0$ we have the turbulent case.

Equation (1) is strongly nonlinear and a result on the uniqueness and existence of a solution to (1),(4)-(6) has not been found. The main difficulty is that the function $a(y,v_y)$,

(7) $\quad a(y,v_y)=\nu+\varepsilon$

is not differentiable with respect to v_y.

Observe that (1) also may be written as

$$\text{(8)}\qquad \begin{aligned} \frac{\partial v}{\partial t} &= A(y,v_y)\frac{\partial^2 v}{\partial y^2} + f(y,v_y) \\ &=(\nu+2\varepsilon)\frac{\partial^2 v}{\partial y^2} + 2\kappa^2 y\left(1-\frac{y}{A}\right)\exp(-2y/A)\,|v_y|\,v_y . \end{aligned}$$

In order to find an approximate solution to (1)-(6), there exist quite a number of different methods to propose, both in the class of difference methods and in the class of finite element methods. Certainely the best choice will not be the Du Fort Frankel scheme. However, the object of this paper is to show how that method behaves on our problem. The main difficulty turns out to be how to define this method for a nonlinear equation of the form (1)-(3).

The present problem was presented to the author by D.Myrhaug who also did most of the computations of this paper.

2. The Du Fort Frankel scheme, linear

For the laminar equation

(9) $\quad v_t=\nu v_{yy}$

the unconditionally stable Du Fort Frankel scheme (DFF) is given by (See Mitchell [2], Lambert 1)

(10) $\quad v_i^{j+1}-v_i^{j-1}=2r\left[v_{i+1}^j-(v_i^{j+1}+v_i^{j-1})+v_{i-1}^j\right],\ r=\nu k/h^2$

and is obtained from the unconditionally unstable Richardson scheme

(11) $\quad v_i^{j+1}-v_i^{j-1}=2r\left[v_{i+1}^j-2v_i^j+v_{i-1}^j\right]$

by replacing $2v_i^j$ by $v_i^{j+1}+v_i^{j-1}$, where h,k are the stepsize in the space- and timedirection.

3. Nonlinear Du Fort Frankel schemes

Let us consider the equation

$$(12)\qquad \frac{\partial v}{\partial t}=\frac{\partial}{\partial y}(a(y,v_y)v_y)\ .$$

Proceeding as in the linear case, we obtain the equation

$$(13)\qquad V_i^{j+1}-V_i^{j-1}=2r\Big\{a(y_{i+\frac12},v^j_{y_{i+\frac12}})(V^j_{i+1}-V^j_i) - a(y_{i-\frac12},v^j_{y_{i-\frac12}})(V^j_i-V^j_{i-1})\Big\}\quad .$$

In going from (11) to (10) we replaced V_i^j by $\frac12(V_i^{j+1}+V_i^{j-1})$. Using this in (13) we have

$$(14)\qquad V_i^{j+1}-V_i^{j-1}= r\Big\{a(y_{i+\frac12},v^j_{y_{i+\frac12}})(2V^j_{i+1}-V^{j+1}_i-V^{j-1}_i) - a(y_{i-\frac12},v^j_{y_{i-\frac12}})(V^{j+1}_i+V^{j-1}_i-2V^j_{i-1})\Big\}\ .$$

The quantities that remain to be sposified are $v^j_{y_{i+\frac12}}$ and $v^j_{y_{i-\frac12}}$.

Since we want an appoximation to v_y at the points $(y_{i-\frac12},t_j)$ and $(y_{i+\frac12},t_j)$ (see Figure 3) a natural second order approximation would be

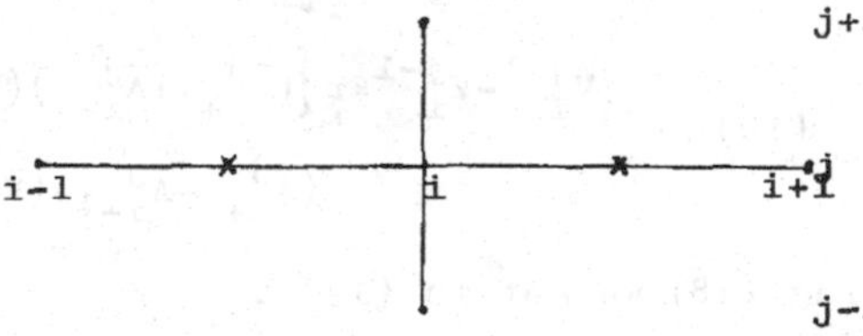

Figure 3

$$(15)\qquad \begin{cases} v^j_{y_{i-\frac12}} \approx (V^j_i-V^j_{i-1})/h \\ v^j_{y_{i+\frac12}} \approx (V^j_{i+1}-V^j_i)/h \end{cases}$$

Define

$$(16)\qquad A_i^j=a(y_{i+\frac12},(V^j_{i+1}-V^j_i)/h)\quad .$$

The resulting nonlinear DFF_1 is then given as

$$(17)\qquad V_i^{j+1}-V_i^{j-1}=2r(A^j_iV^j_{i+1}+A^j_{i-1}V^j_{i-1})-r(A^j_i+A^j_{i-1})(V^{j+1}_i+V^{j-1}_i)\ .$$

<u>Example 1.</u> For water the constants in our problem are

$$\nu=1.14(-6)\ m^2/s,\ \varkappa=0.41\ ,\ \omega=2\pi/30s$$
$$v_o=1.0\ m/s\ ,\ B=300\ s^{-1}\ .$$

(In (3) we replace $|v_y(0,t_j)|$ by $|V_1^j-V_o^j|/h$.) With k=0.05s and h=o.0001 m the results are as shown in Figure 4(Eulers method was used as the starting method.). Infinity is reached when

$$|v| \leq 0.005\ v_o \quad \text{and} \quad |v_y| \leq 0.8 \quad .$$

As the figure shows, instability has already a great impact on the numerical solution for t=9k=0.45s. This could be interpreted as that the approximation in (15) is not the right one to use.

In order to get rid of the instability a tempting way to follow is to insert $\frac{1}{2}(V_i^{j+1}+V_i^{j-1})$ for V_i^j in (15). However, the is nonlinear equations for the unknowns.

A somewhat more simple suggestion is then to use

$$(18) \qquad v^j_{y_{i-\frac{1}{2}}} \text{ and } v^j_{y_{i+\frac{1}{2}}} \approx (V^j_{i+1}-V^j_{i-1})/2h =:\overline{V}^j_i$$

If we assume $a(y,v_y)$ to be sufficiently smooth, a rather tedious calculation shows that the local truncation error is $O(k^2)+O(h^2)+O((k/h)^2)$.

Let $\overline{A}^j_{i+\frac{1}{2}}=a(y_{i+\frac{1}{2}},v^j_{y_{i+\frac{1}{2}}})$. Then (13) may be written as

$$(19) \qquad \begin{aligned} V_i^{j+1}-V_i^{j-1}=r\Big\{&(\overline{A}^j_{i+\frac{1}{2}}+\overline{A}^j_{i-\frac{1}{2}})(V^j_{i+1}-2V^j_i+V^j_{i-1}) \\ &+(\overline{A}^j_{i+\frac{1}{2}}-\overline{A}^j_{i-\frac{1}{2}})(V^j_{i+1}-V^j_{i-1})\Big\} \quad . \end{aligned}$$

Using (18) we get in (19) ,

$$\overline{A}^j_{i+\frac{1}{2}} \pm \overline{A}^j_{i-\frac{1}{2}} \approx a(y_{i+\frac{1}{2}},\overline{V}^j_i) \pm a(y_{i-\frac{1}{2}},\overline{V}^j_i) \quad .$$

Comparing with $v_t=a(y,v_y)v_{yy}+(a(y,v_y))_y v_y$, (18) is by no mean an unnatural choise. Setting

$$B_i^j=a(y_{i+\frac{1}{2}},(V^j_{i+1}-V^j_{i-1})/2h)$$

and changing B_i^j for A_i^j in (17) we obtain the nonlinear DFF_2.

The results of using this scheme on our problem with data as in example 1 are shown in Figure 5. In this case no instabilty occurs and the solution obtained goes towards a periodic function.

In order to find an explination for the difference in these two cases, the following method for (9) may be considered,

$$(20)\qquad v_i^{j+1}-v_i^{j-1}=2r\left\{v_{i+1}^{j}-\left[\theta(v_i^{j+1}+v_i^{j-1})+2(1-\theta)v_{i-1}^{j}\right]+v_{i-1}^{j}\right\}.$$

Here $\theta=0$ gives the Richardsen method (11) and $\theta=1$ the linear DFF in (10).

An easy calculation then gives for the stability,

$\theta<1$: Unconditionally unstable

$\theta\geqslant 1$: stable .

This means that the DFF method is on the boundary of beeing unconditionally unstable, and by using DFF_1 I believe that we pushed DFF over to the unstable side. This was confirmed by the fact that no matter the size of k, we always got an unstable solution.

Let us remark that a further improvement was obtained by smoothing the a-values.

A natural way for getting better results by using the approximation (15) would now be to use a value of $\theta>1$. Results in that direction can be found in Geheler [4] for a linear equation.

4. Method of lines and a semi implicit Runge-Kutta method

As a control, equations (1), (4)-(6) were integrated by using the method of lines and a stiff first order system solver, SIRKUS. By first finding a value Y of y such that (5) can be replaced by $v(y,t)=0$ for y Y, t 0 and choosing a suitable h dividing 0,Y into N+1 intervals of length h, (1) can be approximated by the system

$$(21)\qquad \begin{cases} \dot{V}_i(t)=\frac{1}{h^2}\Big[A_i(t)V_{i+1}(t)-(A_i(t)+A_{i-1}(t))V_i(t) \\ \qquad +A_{i-1}(t)V_{i-1}(t)\Big]\ ,i=1,\dots,N, \\ V_0(t)=v_0\cos(\omega t),\ V_{N+1}(t)=0.0\ . \end{cases}$$

where

$$(22)\qquad \begin{cases} A_i(t)=a((i+\tfrac{1}{2})h,(V_{i+1}(t)-V_i(t))/h) \\ V_i(t)\ v(ih,t)\ ,\ i=0,\dots,N+1. \end{cases}$$

The first order system (21), written as

$$(23)\qquad \vec{V}(t)=\vec{f}(t,\vec{V}(t))=A(\vec{V}(t))\vec{V}(t)+\vec{b}(t,\vec{V}(t))$$

was then integrated by using the program SIRKUS, based on the semi implicit Runge-Kutta algorithm

$$\vec{g}_1^j=\vec{f}(t_j+\gamma k,\vec{V}^j+\gamma k\vec{g}_1^j)$$
$$\vec{g}_2^j=\vec{f}(t_j+9(1-3\gamma)k,\vec{V}^j+(9-28\gamma)k\vec{g}_1^j+\gamma k\vec{g}_2^j)$$
$$\vec{V}^{j+1}=\vec{V}^j+k\left[(43+10\gamma)\vec{g}_1^j+(19-10\gamma)\vec{g}_2^j\right]/62$$
$$\vec{g}_1^{j+1}=\vec{f}(t_{j+1}+\gamma k,\vec{V}^{j+1}+\gamma k\vec{g}_1^{j+1})$$
$$\overrightarrow{(\text{Local error})}_{j+1}=k\left[(28-80\gamma)\vec{g}_1^j-(28-18\gamma)\vec{g}_2^j+62\gamma\vec{g}_1^{j+1}\right]/93$$
$$\vec{V}^j\approx\vec{V}(t_j)=\left[V_1(t_j),\ldots,V_N(t_j)\right]^T$$
$$\gamma=1-\sqrt{2}/2\quad,$$

(See Nørsett [3]).

The absolute local error tolerance was set to 0.1 and starting-stepsize in t-direction to 0.05s. Let us remark that the Jacobian of (23) is tridiagonal and causes a rapid integration of (23) using SIRKUS. The results of this run are as in Figure 6 and shows resonable numbers and no stability problems.

<u>References</u>

[1] Lambert, J.D.; "Variable Coefficient Multistep Methods for Ordinary Differential Equations applied to Parabolic Differential Equations". Topics in Numerical Analysis II, Ed. John Miller. Academic Press, 1974.

[2] Mitchell, A.R.; "Computational Methods in Partial Differential Equations". John Wiley & Sons, 1969.

[3] Nørsett, S.P.; "Semi Explicit Runge-Kutta Methods". Report No. 6/74, Dept. of Math., The University of Trondheim, Norway.

[4] Geheler, E.; "Entwicklung nach Eigenvektoren beim Verfahren von Du Fort Frankel". ZAMM 55, T238-T240 (1975).

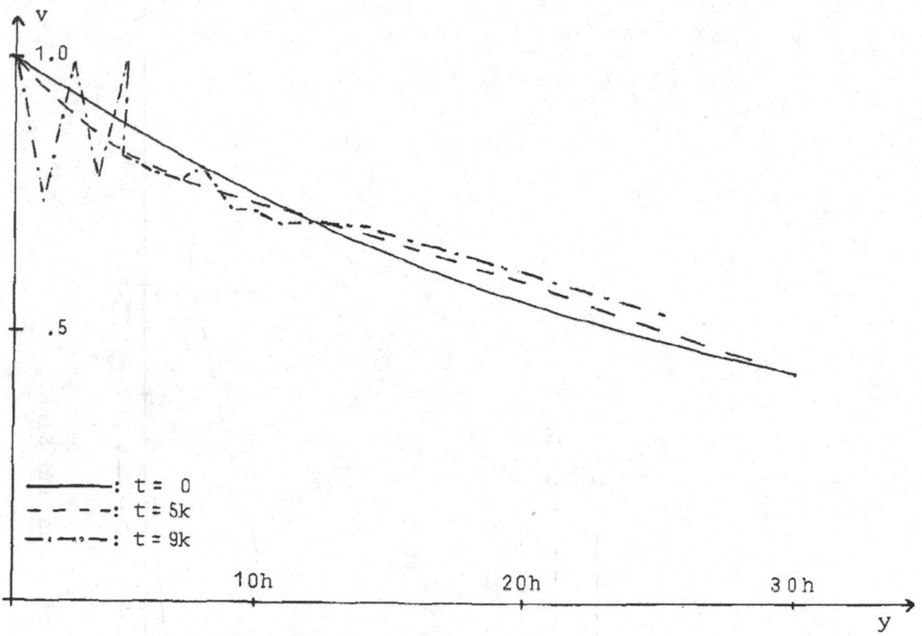

Figure 4. DFF_1 as in example 1

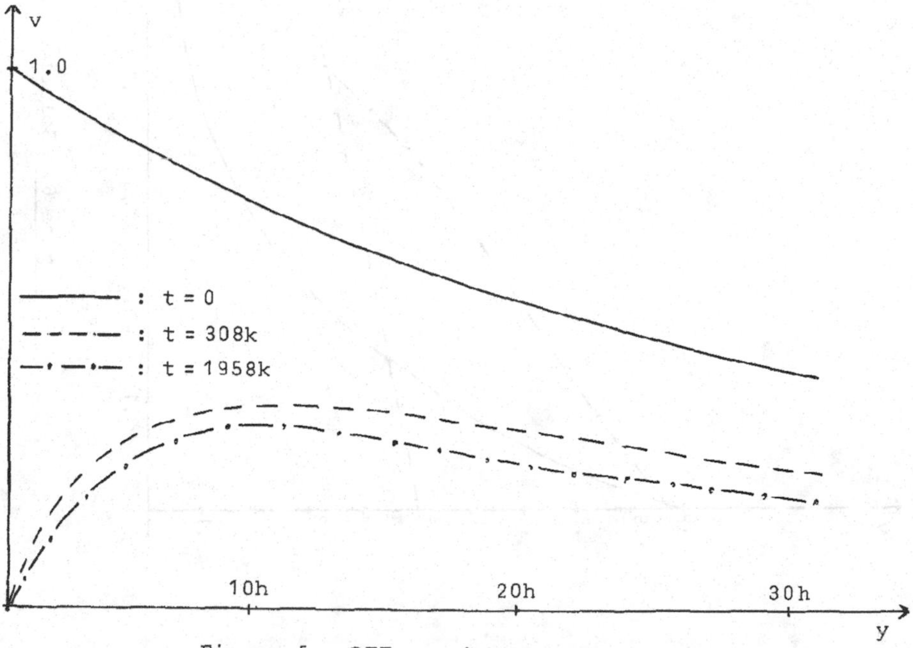

Figure 5. DFF_2 as in example 1

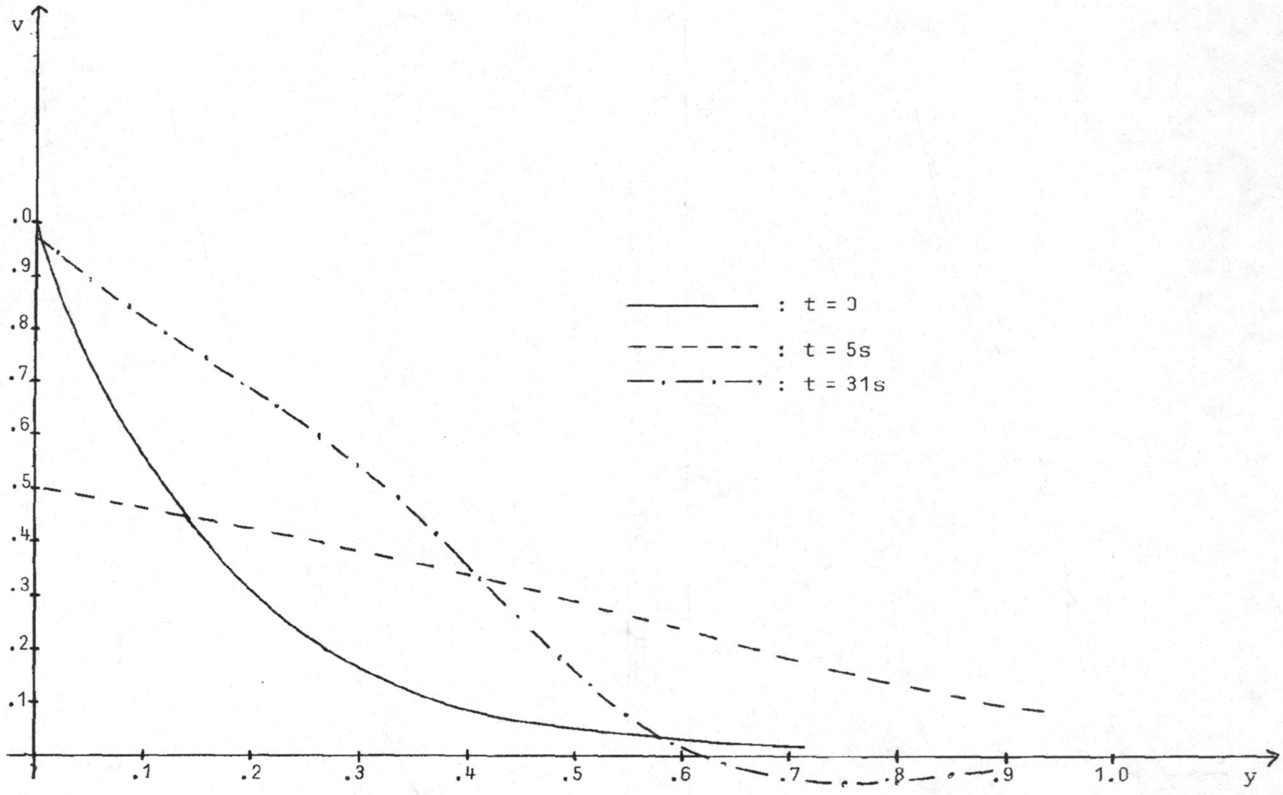

Figure 6. Solution by method of lines and SIRKUS

ON THE NUMERICAL SOLUTION OF NONLINEAR AND FUNCTIONAL DIFFERENTIAL EQUATIONS WITH THE TAU METHOD

E. L. Ortiz

1.-INTRODUCTION. In this paper we describe the application of two methods for the numerical solution of differential equations: the Tau Method and the Method of Condensation. Both are based on a systematic use of the idea of best uniform approximation of functions - implicitely defined by a differential equation, in our case - by polynomial or rational functions, or by piecewise polynomial or rational functions, in their segmented forms.

The Tau Method is briefly described and its use exemplified on the numerical solution of a model differential equation for a singular perturbation problem (linear case); a nonlinear differential equation and a model functional differential equation of retarded type. For the latter we give, both, polynomial and rational approximations. The Method of Condensation is also briefly described and exemplified on a model partial differential equation and the approximation results compared with those given by a collocation method.

Both methods can be described in the framework of projection methods. In the case of the Tau Method the restriction operator is not necessarily linear. In this framework, the Method of Condensation can be given an appealing geometrical interpretation, which we discuss. Finally we make some comments on the fact that, for approximations of a given degree, the Method of Condensation gives more accurate results than collocation or other weighted residuals methods.

2.-A SINGULAR PERTURBATION MODEL PROBLEM AND THE TAU METHOD. The Tau Method. Let D be a linear differential operator

$$D = \sum_{i=0}^{\nu} p_i(x)\,\frac{d^i}{dx^i}\ , \qquad , \quad p_i(x) \in \mathcal{P}_{q(i)}\ , \ x \in J;\ (1)$$

where the coefficients $p_i(x)$ are polynomials (the same arguments apply if they are rational functions) of degree $q_i(x)$, ν is the order of D and J is a finite interval.

Let be given the differential equation:

Problem 1 $\quad\begin{cases} D\,y(x) = F(x) & ,\ x \in J, \\ (f_j, y) = \sigma_j & ,\ j = 1(1)\nu, \end{cases}$

where f_j are functionals such that (f_j,y) describe the supplementary conditions of our problem($(f_j,y)=y^{(j)}$, in the case of an initial value problem).

The Tau approximate solution of Problem 1 is the exact polynomial solution of degree n of the associated perturbed problem:

Problem 2 $\quad\begin{cases} D\,y_n^*(x) = H_n(x) & ,\ x \in J, \\ (f_j, y_n^*) = \sigma_j & ,\ j = 1(1)\nu, \end{cases}$

where $H_n(x)$ is the best uniform approximation of degree m = m(n) (or a sufficiently close approximation of the best) of the right hand side $F(x)$ of Problem 1, This approximation satisfies the same constrains as y. In particular, if $F(x) = 0$, $H_n(x)$ is a Chebyshev polynomial of the first kind defined in J, or a linear combination $\xi_0 T_m(x) + \dots + \xi_k T_{m-k}(x)$ of them. The free parameters ξ_i are adjusted in such a way that the supplementary conditions (and maybe some extra conditions, as we will see immediately) are satisfied.

The differential operator D maps polynomials of degree r into polynomials of degree g. When r runs in the set $N = 0,1,2,\dots,$ g runs in a set of indices $M = N - S$, where S is a finite (or empty) set; the number of elements in S is indicated by s, $s \geq 0$. In the case of a linear differential operator, s is bounded by $\nu + h$, where h is the maximum difference between the degree of p_i in (1) (which rises r) and the order of differentiation i (which lowers r); h is called the height of D (for detailed proofs see ORTIZ [17,18,20]). If, for instance, $D = x^2y' + y$, then there is no polynomial of degree $g = 1$, therefore $S=[1]$.

Let $U = [u_i]$, $i \in N$, be a basis for the space of polynomials $\mathbb{P}$, $\mathbb{R}_S=$ span $[u_i]$, $i \in S$, and $\mathbb{K} = \mathbb{P} - \mathbb{R}_S$. It is clear that D maps $\mathbb{P} \leftrightarrow \mathbb{K} \subseteq \mathbb{P}$, as no polynomial of degree $j \in S$ is the D-image of a polynomial. $\mathbb{R}_S$ is called the subspace of residuals of D. It should be noted that unless S is empty, D produces a collapse of dimension when applied to $\mathbb{P}$.

Let us assume now that with the operator D, a basis $Q = [q_i(x)]$ is given in $\mathbb{K}$, such that $D\, q_i(x) = u_i(x) + r_i(x)$, $i \in M$, $r_i(x) \in \mathbb{R}_S$. Q is called the Lanczos sequence of canonical polynomials associated with D, and r_i the residual polynomial of q_i.

Remark. The elements of Q are classes of equivalence of polynomials, modulo the subspace generated by the exact polynomial solutions of $Dy=0$. This is, however, a technical point which will not be enlarged here (see [18]).

Because of the linearity of D, the solution of Problem 2 is immediately obtained once Q is known: since $H_n = \sum_{i=0}^{m} c_i\, u_i(x)$, we have

$$y_n^*(x) = \sum_{i=0}^{m} c_i\, q_i(x), \text{ for } i \in M,$$

which we can write symbolically as:

$$y_n^*(x) = H_n(\, q_i\,),$$

to emphazise the fact that $y_n^*(x)$ is obtained from H_n by a change of argument which leaves the coefficients invariant.

It is clear now why the number of free parameters ξ_i may, in some cases, be larger than the number ν of supplementary conditions in Problem 1: These extra parameters must be used to account for the collapse in dimensionality caused by D when applied to $\mathbb{P}$; that is, for the fact that some items $u_i \in U$ in the expression of H_n cannot be matched with a D q_i, since q_i remains undefined when $i \in S$. However, they can be matched by a linear combination of all residuals r_i, $i \leq m$, $i \in M$, in H_n. This extra condition (which can equally well be expressed as the cancellation of the coefficients of u_i in H_n) adds s equations to the ν required by the supplementary conditions. It should be pointed out that, in practical applications, $\nu + s$ is a very small number.

Therefore, the construction of the Tau approximant y_n^* involves: i) the generation of the sequence Q, and ii) the inversion of a system of ν+s linear algebraic equations. The first problem can be solved with great simplicity using a recursive algebraic construction described by ORTIZ[17,18], the use of which will be exemplified here.

Remark. The Tau approximant y_n^* is exact of order n in the sense that if the exact solution y of Problem 1 is itself a polynomial of degree n, the Tau Method will reproduce it.

Numerical example 1. Let us consider the model singular perturbation problem:

Problem 1 $\begin{cases} D\,y(x) = y''(x) - 60(x-\tfrac{1}{2})y'(x) = 0, \\ y(0) = 1,\ y(1) = 3,\ x \in [0,1]; \end{cases}$

and the Tau problem associated with it:

Problem 2
$$\begin{cases} D\,y_n^*(x) = y_n^{*\prime\prime}(x) - 60(x-\tfrac{1}{2})y_n^{*\prime}(x) = H_n(x), \\ y_n^*(0) = 1,\ y_n^*(1) = 3,\ x \in [0,1]. \end{cases} \quad (2)$$

Let us take $U = [x^i]$, $i \in N$; in order to construct Q, we first apply D to $u_i = x^i$ and obtain the form of the generating polynomial:

$$\mathbb{P}_n(x) = -60nx^n + 30nx^{n-1} + n(n-1)x^{n-2}, \quad (3)$$

from which we get a recursive expresion for the elements of Q:

$$\text{for } n > 1, \quad q_n(x) = -[\,x^n - 30n\,q_{n-1}(x) - n(n-1)\,q_{n-2}(x)]/60n. \quad (4)$$

From (3) it follows that (for n = 0) the constant $A.x^0$ is an exact polynomial solution of Problem 1; therefore, $S = [0]$ and $s = 1$. Since the exact polynomial solution $A.x^0$ introduces an extra free parameter A, we only need two ξ_i's. With these three parameters we adjust y_n^* to the given boundary conditions and take care of the fact that q_0 remains undefined.

Figure 1 shows the Tau approximate solution of Problem 1 for n=16.

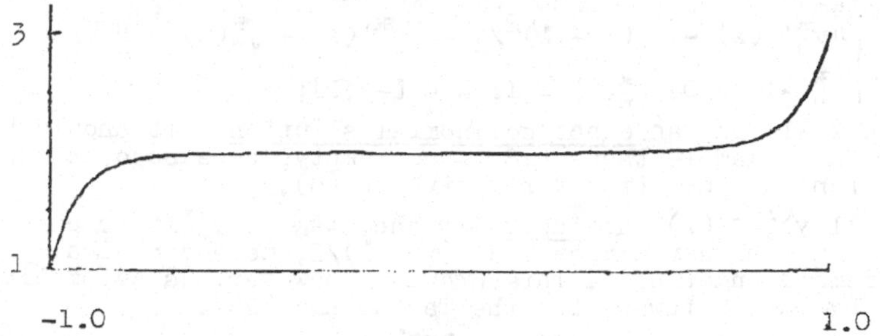

Figure 1. Tau solution of a model singular perturbation problem

Unlike finite difference approximations, this global polynomial solution shows no oscillatory behaviour for n=16. The Tau solution was constructed with the computer software developed at Imperial College for the automation of the Tau Method (see [25]). Time was 112 milliseconds and the computer, CDC 6400.

3.- NONLINEAR DIFFERENTIAL EQUATIONS AND THE TAU METHOD. A boundary value problem. Non linear differential equations , ND y(x) = F(x), can also be treated with the Tau technique (see ORTIZ [21]). Their approximate solution is reduced to a sequence of interrelated lincar Tau problems. The solution of each problem is used to represent the non linearity in the next ,so that the nonlinear differential equation is reduced, in each step, to a linear differential equation with variable coefficients. In this sequential form of the Tau Method we deal with a process

$$\zeta(\,y^{(i)}\,) = y^{(i+1)}, \quad n = 0,1,2,\ldots \quad (5)$$

If the process is contractive , its fixed point is the solution of the nonlinear problem ND y(x) = F(x). The convergence of this process can be measured by the degree of stabilization of the coefficients of succesive approximants, or discussed analytically, in a given interval, in each case. Since each of the intermediate problems is linear, the individual error analysis reduces to that of the Tau Method. Under certain circumstances the magnitude of the parameters ξ_i can be used to estimate the error (see LANCZOS [14,15]); as the ξ_i give an indication of the error in the equation, to obtain information on the error in the solution, some estimation of the inverse operator must be available. In [8,9] we discuss an estimation of the error in the case of simultaneous approximation of the function and derivative (polynomial and rational cases) with this type of approach. The error analysis of the Tau Method can also be constructed using arguments derived from Anselone's theory (see ANSELONE [1]), when applied to collocation problems. A more direct computational approach, which gives a point to point estimation is sketched in 4.- of this paper. It is economical from a computational point of view and can be incorporated easily in the software for the Tau Method.

Numerical example 2. We give now an example of the application of the sequential Tau technique in a concrete situation. A more formidable problem, related to a nonlinear creep problem, where segmented and rational forms are also used, will be considered in a forthcoming paper of HEIMAN and the author.

We consider now the boundary value problem defined by van der Pol's equation:

Problem 1. $\begin{cases} 4y''(x) + 2[y^2(x) - 1]y'(x) + y(x) = 0 \\ y(-1) = 0,\ y(1) = 1,\ x \in [-1,1]. \end{cases}$

As a first approximation we take

$$y^{(0)}(x) = (x + 1)/2,$$

a linear function which satisfies the boundary conditions exactly. We use it to represent the nonliner term in our equation, and formulate the Tau problem

Problem 2. $$\begin{cases} 4y_n^{*}{''}(x) - 2[(x + 1)^2/4 - 1]y_n^{*}{'}(x) + y_n^{*}(x) = H_n(x) \\ y_n^{*}(-1) = 0,\ y_n^{*}(1) = 1,\ x \in [-1,1]; \end{cases} \quad (6)$$

for which we will find an exact polynomial solution. It should be noticed that in this example the set S is not empty, because of the polynomial coefficient of the first derivative in (6).

Let us call $y^{(i-1)}(x)$ the input for the stage i . It is clear that with inputs of a higher degree than $(x + 1)/2$, more canonical polynomials will remain undefined . This problem, however, is automaticaly sorted out by our software for the Tau Method [25].

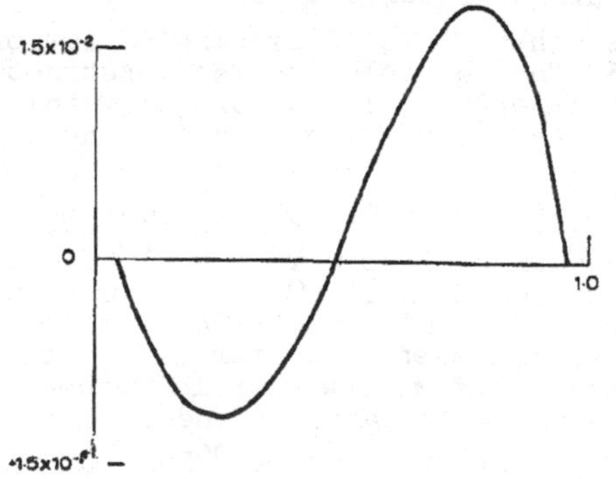

Figure 2. Error graph for a quadratic Tau approximation to van der Pol's equation

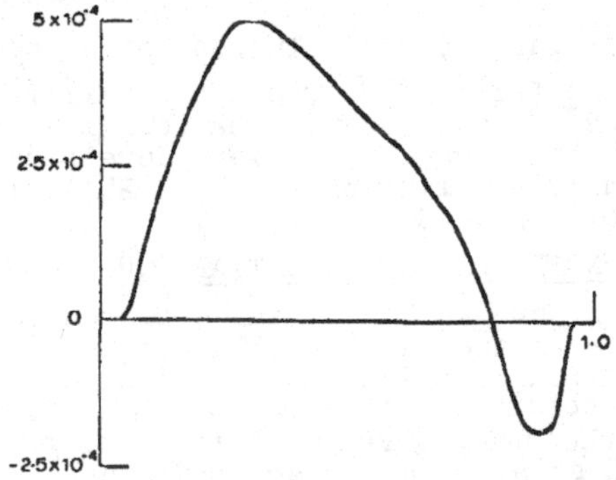

Figure 3. Error graph for a quartic Tau approximation to van der Pol's equation

With the input $y^{(0)}(x)$ we generate the quadratic approximation:

$$y^{(1)}(x) = 0.464\ 285 + 0.500\ 000\ x + 0.035\ 714\ x^2,$$

which exhibits a maximum error bounded by $2.0\ 10^{-2}$; the error function's graph is shown in Fig.2.

A fourth order approximation has a maximum error bounded by $2.5\ 10^{-3}$. Taking as input for a second approximation the quadratic part of this approximation, we find the quartic Tau approximant:

$$y^{(2)}(x) = 0.464\ 973 + 0.538\ 862\ x + 0.047\ 116\ x^2 - 0.038\ 862\ x^3 - 0.012\ 089\ x^4.$$

The maximum error of $y^{(2)}(x)$ remains below $5.0\ 10^{-4}$ and the corresponding error function is represented in Figure 3.

We notice that the difference between succesive Tau approximants, after only two iterations, is

$$\delta_{1,2}(x) = 2.804\ 10^{-3} + 1.308\ 10^{-3}x - 3.457\ 10^{-3}\ x^2,$$

plus the cubic and quartic terms of $y^{(2)}(x)$. The time taken by our CDC 6400 computer system to generate the two approximations was under 100 milliseconds.

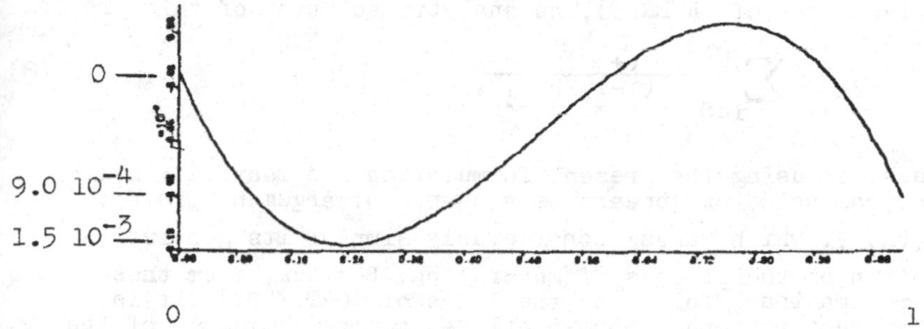

Figure 4. Polynomial Tau approximation to a functional equation, error curve

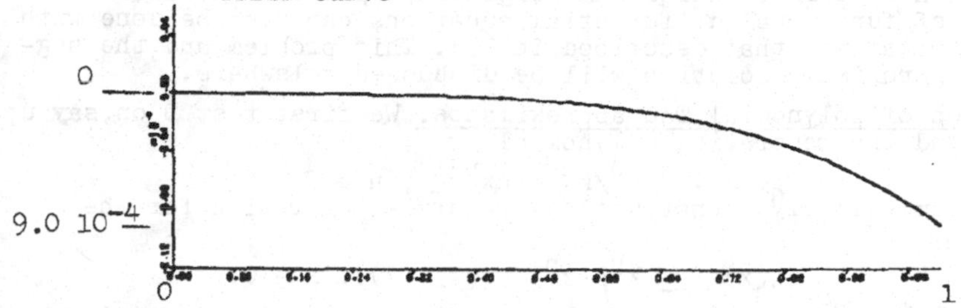

Figure 5. Rational Tau approximation to a functional equation

If instead of powers of x we use, as suggested before, some other basis U, for instance Chebyshev polynomials defined in [-1,1], the sequence of canonical polynomials will generate the elements of sequence (5) in terms of <u>Chebyshev polynomials</u>. For $i = 2$, we get

$$y^{(2)}(x) = 0.483\ 688\ T_0(x) + 0.509\ 715\ T_1(x) + 0.017\ 513\ T_2(x) - 0.009\ 715\ T_3(x) - 0.001\ 511\ T_4(x).$$

<u>Remark.</u> This form of the Tau Method is similar to the Lanczos selected

point method (LANCZOS[14,15]) or Chebyshev orthogonal collocation methods derived from it (see CLENSHAW and NORTON [4]). Indeed the results obtained with step two of this iterative technique reproduce (up to $2.0\ 10^{-5}$) the results of 11 iterations of Chebyshev collocation (see NORTON[16]).

4.- FUNCTIONAL DIFFERENTIAL EQUATIONS AND THE TAU METHOD. A functional differential equation of retarded type. Fox, Mayers, Ockendon and Tayler [7], have considered the application of Lanczos' original formulation of the Tau Method to find approximate solutions of a functional differential equation related to the mechanical behaviour of overhead wires for locomotive collectors . . The equation they considered is of the form $y'(x) = a\, y'(x/k) + b\, y(x)$, where a and b are constants. For this equation they give extensive results. They use, as advocated by Fox in other work, an integrated form of the equation to simplify the manipulation of Chebyshev series they use to solve approximatively this equation.

We will apply here the formulation of the Tau Method just discussed, to treat numerically a similar problem:

Problem 1 $\quad \{ FD\, y(x) = y'(x) \pm y(x/k) = 0,\ y(0) = 1,\ x \in [0,1], k > 1,$ (7)

(the case of infinite lag). This particular problem has been discussed, with other methods, by FELDSTEIN and GRAFTON[6] and is also reviewed in the comprehensive paper of CRYER[5].The analytic solution of this problem is, for k = 2,

$$\sum_{i=0}^{\infty} \frac{(\mp x)^i}{2^{i(i-1)/2}\, i!} \qquad (8)$$

The advantages of using the present formulation are manyfold: here again the numerical solution appears as a change of arguments in H_n: $y_n^*(x) = H_n(\, q_i(x)\,)$, which makes considerably simpler its generation and the automation of the process of generation. Besides, from those polynomial approximations, following the lines of ORTIZ[21], it is possible to construct rational approximations to the solution of the functional equation (7).

The extension of this technique to eigenvalue problems defined by certain types of functional differential equations can also be done with a technique similar to that described in [3]. This problem and the segmentation of approximate solution will be discussed elswhere.

Construction of polynomial Tau approximants. We first test D on, say $u_i = x^i$, and find the generating polynomial

$$\mathbb{P}_n(x) = \pm\, x^n/k^n + nx^{n-1} \ ,\ n \geq 0,$$

from which we immediately construct the recursive expresion for the elements of Q:

$$q_n(x) = \pm\, k^n[\, x^n - nq_{n-1}(x)],\ n \geq 0.$$

The set S is empty. As indicated before, solutions in terms of other types of polynomials can be obtained with a different choice of U([19]).

With (7) we associate

Problem 2
$$\begin{cases} DF\ y_n^*(x) = y_n^{*\prime}(x) \pm y_n^*(x/k) = H_n(x), \\ y_n^*(0) = 1,\ x \in [0,1],\ k > 1. \end{cases}$$

As $H_n(x)$ we take the exact best uniform approximation of the right hand side of (7) by a polynomial of degree n, and with a non zero uniform norm, so that this constrained approximation satisfies the given initial condition.

As $T_n^*(x) = c_0^{(n)} + \ldots + c_n^{(n)}x^n$, $y_n^*(x) = H_n(q_i(x))$ has the form

$$y_n^*(x) = \xi \sum_{i=0}^{n} c_i^{(n)} q_i(x).$$

Taking into account the initial condition, we get

$$y_n^*(x) = \sum_{i=0}^{n} c_i^{(n)} q_i(x) \Big/ \sum_{i=0}^{n} c_i^{(n)} q_i(0).$$

We will discuss numerically the two forms of (7):

i) $y'(x) + y(x/k) = 0, \quad y(0) = 1, \; x \in [0,1], \; k > 1.$

In this case, say, for a quadratic Tau approximation, we get

$$y_{2,k}^*(x) = [8k^2x^2 - 8kx(2k^2+1) + 8k(2k^2+1) + 1]/[8k(2k^2+1) + 1].$$

For $k = 2$,

$$y_{2,2}^*(x) = (32x^2 - 144x + 145)/145,$$

which, at the end point of the interval, takes the value

$$y_{2,2}^*(1) = 0.227\,586;$$

with an error

$$e_{2,2}^*(1) = 2.223\;10^{-3}$$

The approximating value at this particular point of the interval can be improved with the choice (see [21]) $H_n(x) = \xi P_n(x)$, where $P_n(x)$ is the Legendre polynomial of order n defined in the interval $0 \leq x \leq 1$.

With such perturbation we get:

$$\tilde{y}_{2,k}^*(x) = [6k^2x^2 - 6kx(2k^2+1) + 6k(2k^2+1) + 1]/[6k(2k^2+1) + 1]$$

or, for $k = 2$,

$$\tilde{y}_{2,2}^*(x) = (24x^2 - 108x + 109)/109,$$

For $x = 1$,

$$\tilde{y}_{2,2}^*(1) = 0.\,229\,357\;, \text{ and } \tilde{e}_{2,2}^*(1) = 4.518\;10^{-4}.$$

It follows from (8) that $y(x)$ has a first zero $x = \alpha$ in $[0,1.7]$. If we use our quadratic approximation $\tilde{y}_{2,2}^*(x)$ to estimate it, we get $\alpha_{2,2}^* = 1.521$. Replaced in (8) it gives $y(1.521) = -1.312\;10^{-2}$. We have used the Tau approximation **outside** its range of definition; a slightly better result is obtained if the Tau approximant is constructed in the interval $[1,2]$.

ii) $y'(x) - y(x/k) = 0\;, \quad y(0) = 1, \; x \in [0,1], \; k > 1$ (9)

For a quadratic approximation with $H_n(x) = \xi T_2^*(x)$ we find, as before,

$$y_{2,k}^*(x) = [-8k^2x^2 - 8kx(2k^2 - 1) - 8k(2k^2-1) - 1]/[-8k(2k^2-1)-1].$$

If we take $k = 2$,

$$y_{2,2}^*(x) = (32x^2 + 112x + 111)/111,$$

with an error of $2.580\;10^{-2}$ at $x = 1$. An approximation constructed taking $k = 2$, $n = 2$, and a Legendre perturbation term is:

$$\tilde{y}_{2,k}^*(x) = [-6k^2x^2 - 6k(2k^2-1) - 6k(2k^2-1) - 1]/[-6k(2k^2-1) - 1]\;, \quad (10)$$

and, for $k = 2$

$$\tilde{y}_{2,2}^*(x) = (24x^2 + 84x + 85)/85$$

The error in $x = 1$ is now $9.043\;10^{-4}$. The error curve is shown in Fig.4.

Construction of Tau rational approximants. We will now use our accurate end point approximation (10) as the basis for the construction of a rational Tau approximant to the functional differential equation (9). Instead of the interval [0,1] we now use $J_E = [0,1/E]$ as the interval in which a solution of (9) is sought. $E \geq 1$ and $J_E = J$ if $E = 1$. The Tau solution in J_E, for $n = 2$, takes the form

$$\tilde{y}^*_{2,2}(x,E) = [-6E^2k^2x^2 - 6Ekx(2Ek^2-1) - 6Ek(2k^2-1) - 1]/[-6Ek(2k^2-1) -1].$$

If we now identify $1/E$ and x, i.e., if we always compute the approximation at the <u>end point</u> of the interval J_E, we get for (9) the <u>rational Tau approximation</u>

$$\tilde{y}^*_{2,2}(x,x) = [-x^2(6k^2-6k+1) - 6kx(2k^2-1) - 12k^3]/[-x^2 + 6kx - 12k^3],$$

In particular, for $k = 2$:

$$\tilde{y}^*_{2,2}(x,x) = \frac{13x^2 + 84x + 96}{x^2 - 12x + 96}. \qquad (11)$$

For $x = 1$ the Tau rational approximation (11) has an error of $9.043\ 10^{-4}$, same as (10) from which it was constructed (case $E = 1$); the difference between the rational and the polynomial approximations becomes more significant as we move away from the point $x = 1$, as shown by Table I; see also Fig.5 .

TABLE I

Error of :	$y^*_{2,2}(x)$	$\tilde{y}^*_{2,2}(x)$	$\tilde{y}^*_{2,2}(x,x)$
$x =$	$e_{2,2}$	$\tilde{e}_{2,2}$	rat. $\tilde{e}_{2,2}$
0.50	$1.143\ 10^{-2}$	$4.392\ 10^{-4}$	$4.816\ 10^{-5}$
0.25	$4.238\ 10^{-3}$	$1.326\ 10^{-3}$	$8.233\ 10^{-5}$
0.20	$3.165\ 10^{-3}$	$1.226\ 10^{-3}$	$1.115\ 10^{-6}$
0.10	$1.263\ 10^{-3}$	$8.737\ 10^{-4}$	$6.722\ 10^{-8}$

<u>Direct estimation of the error</u>. Let us call $e_n(x) = y^*_n(x) - y(x)$, the error function. Since FD is <u>linear</u>, it follows that $e_n(x)$ satisfies the same equation as $y^*_n(x)$, but with homogeneous conditions:

$$\{ \text{FD}\ e_n(x) = H_n(x)\ , \ e_n(0) = 0\ , \ x \in [0,1].$$

We apply to this equation the Tau Method again in order to estimate $e_n(x)$. To this end we introduce the associated <u>Tau problem</u>

$$\{ \text{FD}\ e^*_n(x) = H_n(x) + \underline{H}_m(x), \quad e^*_n(0) = 0, \ x \in [0,1],$$

where $m \geq n$.

In the case of (9), for $n = 2$, $k = 2$, we shall only take $m = 3$ and solve the functional differential equation

$$\{ \text{FD}\ e^*_2(x) = T^*_2(x)/111 + \underline{\tau}\, T^*_3(x), \quad e^*_2(0) = 0, \ x \in [0,1],$$

from which we get

$$e^*_2(x) = -0.022\ 154\ 9\ x^3 + 0.033\ 943\ 1\ x^2 - 0.008\ 936\ 1\ x\ .$$

For $x = 1$, this expresion gives $e^*_2(1) = 2.582\ 10^{-2}$, the exact result was $2.580\ 10^{-2}$.

<u>Remark</u>. This technique for the practical estimation of the error can be easily incorporated **in** the software for the computer implementation of the Tau Method. It has a further advantage, with it we can estimate the points where the <u>segmentation</u> of our solution is most desirable, and then construct an <u>adaptive Tau Method</u>, which is described in a separate note.

<u>Functional equations, differential equations and the Tau Method.</u> A differential equation can be recovered from its sequence of canonical polynomials: once the form of the generating polynomial $\mathbb{P}_n(x)$ is obtained by interpolation over a suitable number of elements of Q, it is only a matter of equating coefficients of n to find the form of D.

In the case of the functional differential equation we are consiring, the expresion of $\mathbb{P}_n(x)$ contains not only powers of n, but the exponential factor k^{-n}. This a distinctive feature of these functional equations. Considering polynomial approximations to the exponential factor we can associate with the functional equation a <u>family of differential equations</u>, the elements of which can be discussed as ordinary Tau problems.

<u>5.- PARTIAL DIFFERENTIAL EQUATIONS AND THE METHOD OF CONDENSATION</u>. The Tau method can be extended to the case of partial differential equations following essentially the same arguments given in 2.- of this paper. For a more detailed treatment see [26,27]. We shall not deal with those extensions here, but with the so-called Method of Condensation, proposed by HEIMAN and ORTIZ [11-13] in connection with the numerical approximation of the velocity field of a solid in the process of extrusion.

Let $\epsilon > 0$ be an admissible error bound, z_n a given approximate solution of a partial differential equation, and $z = z(x,y, \ldots ,t)$ the exact solution of such equation. If η is the error between z and z_n in the domain in which the solution is required, we will assume, further, that ϵ is small compared with η. The compound error $\eta + \epsilon$ may, for instance, be the error of a graph plotting device.

We will also assume (which is not essential) that z_n has been obtained by means of a projection method, say, by a weighted residuals technique, and that it is a polynomial (or a piecewise polynomial defined in a certain finite element) belonging to a certain subspace of polynomials $\mathcal{P}_n$.

The Method of Condensation attempts to find a projection z_n^ϵ of z_n on a subspace $\mathcal{P}_m$ of $\mathcal{P}_n$ $(m \leq n)$, such that m is as small as possible while the maximum absolute value of the difference between z_n^ϵ and z_n remains bounded by ϵ in the domain in which the approximate solution is sought.

From a numerical point of view, the Method of Condensation can be regarded as a technique for the <u>acceleration of the convergence</u> of the approximate solution z_n, in the sense that roughly the same error is obtained with a polynomial of a <u>lower</u> degree. If we identify $\mathcal{P}_n$ with $\mathbb{R}_{n+1}$, where the vectors $\underline{a}_n \in \mathbb{R}_{n+1}$ are the coefficients of polynomials z_n, this technique can be extended to discrete variable methods.

From a computational point of view, the Method of Condensation implements a <u>reduction of the complexity in the evaluation</u> of z_n, within an error of a given amplitude ϵ. This reduction can be quite significant (a 60% in the example of pp. 337-8 of [23]).

Clearly the efficiency of this process depends on the smoothness of z_n; however, the efficiency of the process can be <u>estimated in advance</u> from the coefficients of z_n. This is exploited in the software for the Method of Condensation developed at Imperial College [2].

The Method of Condensation can be stated in the framework of projection methods, where it appears as a re-projection technique. Let V, S be linear normed spaces, $\lambda = [\lambda_i]$, $i \in N$, be a coordinate system for V, $\mathcal{P}_n$ and S_n finite dimensional approximating subspaces of V and S respectively, $p_n: \mathcal{P}_n \mapsto V$ and $r_n: S \mapsto S_n$ are linear and continuous (r_n may be nonlinear, as in the case of the Tau Method, where the projection is defined by the operator of best uniform approximation by polynomials, with constrains). The element $z \in V$ is the exact solution of the problem $Dz = F$, $D : V \mapsto S$, and z_n is the approximate solution of $Dz = F$ in the projection sense, that is, $p_n z_n$ is mapped by D into an element such that its projection on S_n coincides with that of F: $r_n D p_n z_n = r_n F$.

The element z_n is found from F_n by means of a numerical method which we indicate with Δ_n^{-1}. We now take a projection of the polynomial z_n: $C^{\varepsilon}_{n,m}: \mathcal{P}_n \mapsto \mathcal{P}_m$ ($m = n, n+1, \ldots$) until the constrain $\| z_n - z_n \| < \varepsilon$ ceases to be satisfied. We take as the condensed solution of $Dz = F$ the last element of that sequence for which the constrain is satisfied. This projection is implemented by means of a technique of near-best approximation, in our computer implementation, projection is made on a product Chebyshev basis defined in the domain in which the solution is required.

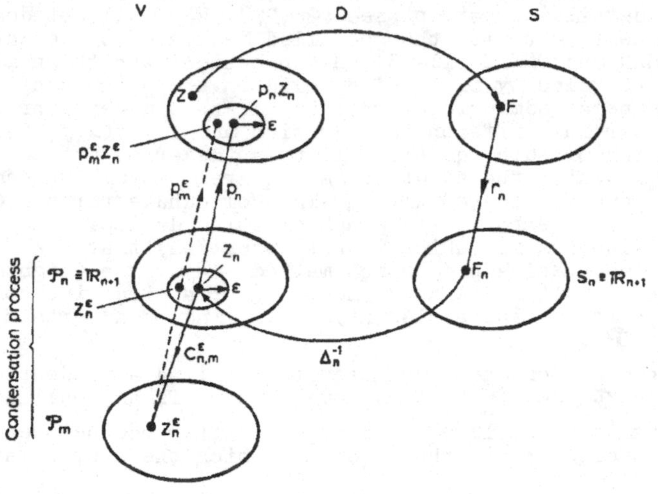

Figure 6. Condensation technique in relation to projection methods

Numerical example 4. The table below shows numerical results obtained comparing in a grid of points (x_i, y_i), $i = 0(0.25)1$, the analytic solution of the model elliptic problem $\nabla^2 f(x,y) = -2$, $f|_{\overline{D}} = 0$, for $D = [x,y: 0 \leq |x|, |y| \leq 1]$ (torsion problem of Saint-Venant), with approximate solutions obtained applying the method of Chebyshev orthogonal collocation (approximate solution of order 4), and applying the Method of Condensation to an unsophisticated least squares approximation of order 6 and then condensed to order 4. The condensation solution has a maximum error of 0.89 %, whereas the collocation solution exhibits a maximum error of 11.64 %.

The Chebyshev collocation method may fail to produce a better answer in a subspace of polynomials of a fixed degree (4 in our case) because the Chebyshev perturbation that is placed in the right hand side of the equation in order to solve it as an interpolation problem, does not share, with the well behaved one dimensional case, the same near best approximating properties. Notice that there are variable coefficients in the expresion of the Chebyshev product (see [13]).

In the one dimensional case the Condensation Method will reduce to the Tau Method and be equivalent to Lanczos' condensation.

TABLE II

y \ x =	0	0.25	0.50	0.75
y=0	0.589 0,585 0.600	0.558 0.553 0.562	0.459 0.456 0.450	0.280 0.279 0.262
0.25		0.528 0.524 0.530	0.436 0.434 0.431	0.267 0.266 0.258
0.50			0.362 0.362 0.366	0.226 0.225 0.234
0.75				0.146 0.146 0.163

Numerical results in this table:

first line : analytic sol.
second line: condensed sol.
third line : collocation sol.

An application of the Method of Condensation to a less trivial exemple can be found in H.HEIMAN [10].

R E F E R E N C E S

[1] Anselone, P.M. (1971). Collectively Compact Operators Approximation Theory, Prentice-Hall, New Jersey.

[2] Arce, G., Helman, H. and Ortiz, E.L. (1977). Software for numerical condensation in two variables. I.C. Res. Report.

[3] Chavez, T. and Ortiz, E.L. (1968). On the numerical solution of two point boundary valuc problems for linear differential equations, Z.angew.Math.Mech., 48, 415-418.

[4] Clenshaw, C.W. and Norton, H.J. (1963). The solution of nonlinear ordinary differential equations in Chebyshev series, The Computer J., 6, 88-92.

[5] Cryer, C. (1972). Numerical methods for functional differential equations, in : Delay and Functional Differential Equations and Their Applications, K.Schmitt, ed. Aacdemic Press, New York

[6] Feldstein, A. and Grafton, C.K. (1968). Experimental mathematics: an application to retarded ordinary differential equations, Proc. 23rd Nat. Conf. Assoc. Comp.Mach., 67-71.

[7] Fox, L; Mayers, D.F.; Ockendon, J.R. and Tayler, A.B. (1971). On a functional differential equation, J.Inst.Maths. Applic., 8, 271-307.

[8] Freilich, J. and Ortiz, E.L. (1977). $C^{(1)}$polynomial approximation with the Tau Method. I.C. Res. Report.

[9] Freilich, J. and Ortiz, E.L. (1977). End point simultaneous approximation with the Tau Method, I.C. Res. Report.

[10] Helman, H. (1976). Hydrostatic extrusion of bimetallic composites, Ph.D. Thesis, Imperial College, University of London.

[11] Helman, H. and Ortiz, E.L. (1975). A new method for the numerical solution of partial differential equations based on condensation in several variables., Proc. Fifth Canadian Congress of Applied Mech., Fredericton, 669-670.

[12] Helman, H. and Ortiz, E.L. (1977). The Method of Condensation, Proc. Proc. Int. Symposium on Innovative Numerical Analysis in Applied Engineering Science, Versailles, Sppl.20-22.

[13] Helman, H. and Ortiz, E.L. (1977). Partial differential equations and the method of condensation, I.C. Res. Report.

[14] Lanczos, C. (1938). Trigonometric interpolation fo empirical and analytical functions, J.Math.Phys., 17, 129-199.

[15] Lanczos, C. (1956). Applied Analysis, Prentice-Hall, New Jersey.

[16] Norton, H.J. (1964). The iterative solution of non-linear ordinary differential equations in Chebyshev series, The Computer J., 7, 76-85.

[17] Ortiz, E.L. (1964). On the generation of the canonical polynomials associated with certain linear differential operators, I.C. Res. Report.

[18] Ortiz, E.L. (1969). The Tau Method, SIAM J. Numer. Anal., 6, 480-92.

[19] Ortiz, E.L. (1972). A recursive method for the approximate expansion of functions in a series of polynomials, Comp.Phys.Comm., 4, 151-156.

[20] Ortiz, E.L. (1974). Canonical Polynomials in the Lanczos Tau Method, in Studies in Numerical Analysis, ed.:B.K.P.Scaife, Academic Press, New York, 73-93.

[21] Ortiz, E.L. (1975). Sur quelques nouvelles applications de la méthode Tau, in Séminaires IRIA, Analyse et Controle de Systemes, Paris, 247-257.

[22] Ortiz, E.L. (1975). Step by step Tau Method. Part 1: Piecewise polynomial approximations, Comp.and Math. with Appli.,1,381-392.

[23] Ortiz, E.L. (1977). Polynomial condensation in one and several variables with applications, in Topics in Numerical Analysis III, ed.: J.J.H.Miller, Academic Pres, New York,327-360.

[24] Ortiz, E.L. On the numerical approximation of certain types of functional differential equations (to be published).

[25] Ortiz,E.L., Purser, W.F.C.,and Rodriguez L.-Canizares,F.J. (1972). Automation of the Tau method. I.C. Res. Report.

[26] Ortiz, E.L. and Wright,C. Numerical solution of partial differential equations with the Tau method (to be published)

[27] Wright, C.(1977).On the solution of partial differential equations, M.Phil Thesis, Imperial College, University of London.

On the Uniqueness and Stability of Weak Solutions of a Fokker-Planck-Vlasov Equation

Reimund Rautmann

Summary: In [7] the existence of weak solutions of a Fokker-Planck-Vlasov equation is proved. In this paper, with a little more stringent assumption we show the uniqueness of weak solutions and establish a criterion of (asymptotic) stability against local disturbations.-As a consequence, in the case of uniqueness the Galerkin method used in [7] is a constructive one, i.e. the whole sequence of all Galerkin-approximations converges.

1. The Problem

The Fokker-Planck-Vlasov Equation

$$(1.1) \qquad u_t + u_y \cdot z + u_z \cdot z \times B + u_z \cdot K_0 u = \varepsilon u_{zz} \text{ for } t > 0,$$
$$u = u_0 \quad \text{for } t = 0$$

describes how a charge distribution $u(t,y,z) \geq 0$ moves in the 6-dimensional phase space $\Omega = R^3_y \times R^3_z$ (with spatial coordinates $y = (y^1, y^2, y^3)$,

$u_y = (\frac{\partial}{\partial_{y^1}} u, \frac{\partial}{\partial_{y^2}} u, \frac{\partial}{\partial_{y^3}} u)$, and velocity coordinates $z = (z^1, z^2, z^3)$,

$u_z = (\frac{\partial}{\partial_{z^1}} u, \frac{\partial}{\partial_{z^2}} u, \frac{\partial}{\partial_{z^3}} u)$) under the influence of its own Coulomb-force

$K_0 u = (K_0 u)(t,y)$ and of a prescribed magnetic force $z \times B(t,y)$ with the given continuous and bounded vector function $B(t,y)$.

Equation (1.1) differs from the Vlasov-equation on its right side only: for any given diffusion coefficient $\varepsilon > o$, the Laplacean

u_{zz} ($= \sum_{i=1}^{3} \frac{\partial^2}{(\partial z^j)^2} u$) models the diffusion of impuls (e.g. due to the collision of particles) in the velocity-space R^3_z.

2. The Weak Formulation of the Problem

Let D be the class of all real functions having partial derivatives of any order and a compact support in $[o,\infty) \times \Omega$. We set $\Omega_T = [0,T] \times \Omega$ for any $T \in (0,\infty)$. For any classical solution[1)] u by multiplication with a function $e \in D$ and integration by parts we get from (1.1) the formula

$$(2.1\text{ a}) \qquad \int_\Omega u\, e \Big|_0^T = \int_{\Omega_T} \{u \cdot (e_t + e_y \cdot z + e_z \cdot z \times B + e_z \cdot K_0 u) - \varepsilon u_z \cdot e_z\}.$$

Moreover, we require the initial condition

[1)] In addition to the existence and continuity of the partial derivatives in (1.1), we have to impose on u an asymptotic condition which ensures the existence of the Coulomb-force $(K_0 u)(t,y)$.

(2.1 b) $$\lim_{t \searrow 0} |u(t,\cdot) - u_0|_{L^2(\Omega)} = 0.$$

For short, we do not write down the differentials of the variables of integration Let V denote the class of all bounded maps $v : [0,\infty) \to L^2(\Omega)$ with $v \in L^2(\Omega_T)$ and the weak derivatives $\frac{\partial}{\partial z^i} v \in L^2(\Omega_T)$ for any $T \in (0,\infty)$, $i = 1,2,3$. We call "weak solution" of (1.1) any $u \in V$ which for a given initial charge distribution $u_0 \in L^2(\Omega)$ solves (2.1) for all $e \in D$ [2], $T \in (0,\infty)$.

3. On the Uniqueness of Weak Solutions

Our proof will work under the

Assumption 3.1. [3]:

The map $K_0 : v \to K_0 v$ is a bounded linear one of $L^2(\Omega)$ in $L^\infty(R_y^3)$.

First we prove the

Theorem 3.1.: *With the assumption 3.1. for any initial value $u_0 \in L^2(\Omega)$ there is at most one weak solution of the initial value problem (1.1).*

Essentially, our proof is a transposition of the method, by which Serrin in [8] pointed out conditions of uniqueness and stability for Navier-Stokes problems.

[2] or, equivalently, for all elements of a complete orthonormal system in $L^2(\Omega_T)$ which is contained in D. By applying the classical proof of E. Hopf in [2, p.224], the equivalence of the weak solutions in this sense with the weak solutions defined in [7]follows immediately.

[3] This assumption, for example, applies to the y-gradient of the modified Coulomb-potential

$$U(t,y) = \alpha \int_\Omega |y-y'|^{-1} e^{-\beta |y-y'|} c(z) u_\delta(t,y',z)\,dy'dz$$

with the Yukawa-factor $e^{-\beta|y-y'|}$ and the cut-off-function $c(z) \equiv 1$ for $|z| \le \gamma$, $c(z) \equiv 0$ for $|z| > \gamma$ (as ususal in physics), and the (spatial) regularization u_δ, converging to u with $\delta \to 0$ ($|y|,|z|$ Euclidean norm in R^3, $\alpha \in R^1,\beta,\gamma,\delta$ positive constants).

In order to prove the theorem, let $u^1, u^2 \in V$ be two weak solutions of (1.1), i.e. we have the equations

$$(3.1\ a)\qquad \int_\Omega u^i e^i \Big|_0^T = \int_{\Omega_T} \{u^i \cdot (e_t^i + e_y^i \cdot z + e_z^i \cdot z \times B + e_z^i \cdot K_0 u^i) - \varepsilon\, u_z^i \cdot e_z^i\}$$

and

$$(3.1\ b)\qquad \lim_{t \searrow 0} |u^i(t,\cdot) - u_0^i|_{L^2(\Omega)} = 0$$

with $u_0^i \in L^2(\Omega)$, for all $e^i \in D$ and $T \in (0,\infty)$, $i = 1,2$. Because the Sobolev-Space $W_1^2 = W_1^2(\Omega_T)$ of all functions $v \in L^2(\Omega_T)$, which have the weak derivatives

$\frac{\partial}{\partial z^j} v \in L^2(\Omega_T)$ for $j = 1,2,3$, is the closure of D with respect to the norm

$$(3.2)\qquad |v|_H = \{\int_{\Omega_T} (v \cdot v + v_z \cdot v_z)\}^{1/2},$$

there are sequences $(u_k^i) \subset D$ with

$$(3.3)\qquad |u^i - u_k^i|_H \to 0\ .$$

Obviously with the u_k^i in the place of the e^i in (3.1) we cannot go to the limit $k \to \infty$. Therefore we introduce for any $v \in L^2(\Omega_T)$ the (partial) mollification

$$(3.4)\qquad v_\tau(t,y,z) = \int_0^T \int_{R^3} \omega_\tau(t-t', y-y')v(t',y',z)dt'dy'$$

containing the mollifier

$$(3.5)\qquad \omega_\tau(y_*) = \begin{cases} 0 & \text{for } |y_*| \geq \tau \\ \exp\left\{|y_*|^2(|y_*|^2 - \tau^2)^{-1}\right\} & \text{for } |y_*| < \tau \end{cases}$$

with the vector $y_* = (t,y) \in R^4$ and any $\tau > 0$.
Taking into account the well-known properties of the mollification we state the

Corollary 3.1.: Assume $v \in V$, $v_k \in D$ and $\lim_{k\to\infty} |v - v_k|_H \to 0$. Then we have $v_{k\tau} \in D$. The mollifications $v_{k\tau}(t,\cdot)$ converge in $L^2(\Omega)$ to $v_\tau(t,\cdot)$ for any $t \in [o,T]$ and the derivatives $v_{k\tau,t}$, $v_{k\tau,y}$, $v_{k\tau,z}$ in $L^2(\Omega_T)$ to $v_{\tau,t}$, $v_{\tau,y}$, $v_{\tau,z}$, respectively.

A further difficulty stems from the unbounded factor z in (3.1). We can overcome it, if - for the present - we introduce two C_∞-functions p and $q = p \cdot p$, which depend on the variable $z \cdot z = \sum_{j=1}^{3} (z^j)^2$ only and have a compact support.

Owing to the special form of the function q and of the (continuous and bounded) vector function $B = B(t,y)$ we have the

Corollary 3.2.: The relations

(a) $(q v)_z \cdot z \times B = v_z \cdot q z \times B$,

(b) $\int_{\Omega_T} \{(v^1 (q v^2)_z + (q v^1)_z v^2) z \times B\} = 0$

hold for any functions $v, v^1, v^2 \in W_1^2$.

The proof of (a) follows by a short calculation using

(3.6) $$\operatorname{div}_z (q z \times B) = 0.$$

The factor q guarantees the existence of the integral in (b) for any two functions $v^1, v^2 \in W_1^2$. Using (3.6) again and Fubini's and the Gauss theorem, we get (b), at first for smooth functions and then on W_1^2 by the usual approximations.

After this preparation we insert the functions

$e^1 = q \cdot u^2_{k\tau}$ $(= (qu^2_k)_\tau)$ and $e^2 = q \cdot u^1_{k\tau}$ in (3.1.a) with the u^i_k from (3.3.).

These e^i belong to D. Taking into account assumption 3.1. and our corollary 3.2.(a) and 3.1., we can go to the limit $k \to \infty$ in (3.1. a). By adding together the two resulting equations, four terms cancel out in virtue of the

Corollary 3.3.: For any two functions $v^1, v^2 \in L^2(\Omega_T)$, the equations

$\int_{\Omega_T} q \cdot \{v^1 v^2_{\tau,t} + v^1_{\tau,t} v^2\} = 0$ and $\int_{\Omega_T} q \cdot z \cdot \{v^1 v^2_{\tau,y} + v^1_{\tau,y} v^2\}$ hold.

The proof is based on the fact, that the mollifier ω_τ in (3.5.) is an even function of any single variable.

The resulting equation contains derivatives with respect to the variables z^j only, $j = 1,2,3$.

Therefore, we can take the limit $\tau \to 0$. For the left side of the equation we have the formula

$$(3.7)\qquad \lim_{\tau\to 0} \int_\Omega u^1 u^2_\tau \Big|_0^T = \frac{1}{2} \int_\Omega u^1 u^2 \Big|_0^T ,$$

because any solution of (3.1) is weakly continuous in the variable $t \in [0,\infty)$. On the right side of the equation we use the strong convergence of the derivatives $u^i_{\tau,z}$ to u^i_z in $L^2(\Omega_T)$ for $\tau \to 0$. Thus we get from (3.1. a) the equation

$$(3.8)\qquad \int_\Omega qu^1u^2 \Big|_0^T = \int_{\Omega_T}\{u^1(qu^2)_z\, K_0 u^1 + (qu^1)_z u^2 K_0 u^2 - \varepsilon u^1_z (qu^2)_z - \varepsilon (qu^1)_z\, u^2_z\}.$$

With the assumption 3.1., the extension of Lemma 2.1 in [7] from smooth functions to the class W^2_1 leads to the

<u>Corollary 3.4.:</u> For any two functions $u^1,u^2 \in V$ and any p, $p \in D$ or $p \equiv 1$ the equation

$$\int_{\Omega_T} \{(pu^1(pu^2)_z + (pu^1)_z\, pu^2)K_0 u^1\} = 0$$

holds.

By means of this corollary and the obvious equations

$$(p^2u)_z = p(pu)_z + p\,p_z\,u \text{ and } u^1_z(p^2u^2)_z + (p^2u^1)_z\,u^2_z = \begin{cases} 2(pu^1)_z(pu^2)_z \\ -2(p_z)^2u^1u^2 \end{cases}$$

we transform (3.8) into the equation

$$(3.9)\qquad \int_\Omega pu^1pu^2 \Big|_0^T = \begin{cases} \int_{\Omega_T}\{ pu^2(pu^1)_z K_0(u^2-u^1) - 2\varepsilon(pu^1)_z(pu^2)_z\} + \\ + |p_z|_\infty \int_{\Omega_T}\{\frac{p_z}{|p_z|_\infty}\, pu^1u^2 \cdot K_0(u^1+u^2) + 2\varepsilon \frac{(p_z)^2}{|p_z|_\infty}\, u^1u^2\} \end{cases}.$$

Now we take a sequence [4] (p_n) of C_∞-functions, which depend on the variable $z \cdot z$ only and have compact support. We assume $0 \le p_n(z) \le 1$ for all z, $p_n(z) \equiv 1$ for $z \cdot z \le n$ and $|p_{n,z}|_\infty \to 0$ for $n \to \infty$.

Owing to assumption 3.1., the second integral on the right side of (3.9) with $p = p_n$, $n \to \infty$ remains bounded, whereas the first integral on the right and the

[4] The following construction was obtained by G. Hannoschöck, c.p. [1].

integral on the left side converge to the finite value, which corresponds to the case $p \equiv 1$. Therefore, our result is the equation

$$(3.1o)\quad \int_\Omega u^1u^2 \Big|_0^T = \int_{\Omega_T} \{u^2u^1_z\, K_0(u^2-u^1) - 2\varepsilon\, u^1_z u^2_z\}\,.$$

In the special case $u^1 = u^2 = u$ we get

$$(3.11)\qquad \int_\Omega (u(T,\cdot))^2 + 2\,\varepsilon \int_{\Omega_T} (u_z)^2 = \int_\Omega (u_0)^2.$$

This equation shows the continuity of the norm $|u(t,\cdot)|_{L^2(\Omega)}$ in the variable t. Because any solution $u \in V$ of (3.1.) is weakly continous, we have the

<u>Corollary 3.5.:</u> With the assumption 3.1., any weak solution of (3.1.) is strongly continuous in $L^2(\Omega)$.

For two weak solutions $u^1, u^2 \in V$ of (3.1.) we add together the equations (3.11) for u^1,u^2, respectively and subtract twice the equation (3.1o). Taking into account corollary 3.4., we get for the difference $w = u^2 - u^1$ the relation

$$(3.12)\qquad \int_\Omega (w)^2 \Big|_0^T + 2\varepsilon \int_{\Omega_T} (w_z)^2 = 2\int_{\Omega_T} u^2 w_z K_0 w.$$

According to the Cauchy-Schwarz inequality and Cauchy's inequality $2ab \le \frac{1}{2\varepsilon}\, a^2 + 2\varepsilon\; b^2$, the right side in (3.12) has the bound

$$c \int_0^T \int_\Omega (w)^2 + 2\varepsilon \int_{\Omega_T} (w_z)^2$$

with the constants $c = |u_o^2|^2_{L^2(\Omega)} \cdot \frac{c_0^{\,2}}{2\varepsilon}$ and c_0 from $|K_0 v|_\infty \le c_0\, |v|_{L^2(\Omega)}$ (assumption 3.1). Therefore, the continuous function $\varphi(t) = |w(t,\cdot)|^2_{L^2(\Omega)}$ is a solution of the linear integral inequality

$$(3.13)\qquad \varphi(t) \;\le\; \varphi(o) + c\int_0^t \varphi(t')dt'.$$

From this, Gronwalls Lemma leads to the estimate

$$\varphi(t) \le \varphi(o)\cdot e^{c\cdot t} \quad \text{for } t \ge o,$$

which verifies theorem 3.1.

4. A Stability Criterion

Assumption 4.1.: A weak solution u^2 of (1.1) exists, which is bounded almost everywhere, and the difference $w = u^2 - u^1 \in V$ of u^2 and a second weak solution u^1 vanishes identically on the complement of the ball $\mathcal{K} = \{z \mid |z - z_0| < d\}$ *in* the velocity-space R_z^3 (i.e. we have $w(t,y,z) = 0$ for all $t \geq 0$, $y \in R_y^3$ and $z \notin \mathcal{K}$ with a $z_0 \in R_z^3$ and $d > 0$). In addition to assumption 3.1., the map K_0 is a bounded linear one of $L^2(\Omega)$ in $L^2(R_y^3)$, too.

The considerations, which led us to the inequality (3.13), apply in the case of any initial value $t_0 \in [o,\infty)$. With the assumption 4.1., using again the Cauchy-Schwarz inequality and Cauchy's inequality, we get the integral inequality

$$(4.1) \qquad \varphi(t) \leq \varphi(t_0) + c_2 \int_{t_0}^{t} \varphi(t')dt' - \varepsilon \int_{t_0}^{t} \int_{\Omega} (w_z)^2$$

for $\varphi(t) = |w(t,\cdot)|_{L^2(\Omega)}^2$ with the constants

$$(4.2) \qquad c_2 = \frac{c_1^2}{\varepsilon} |u^2|_\infty^2 \quad \text{and } c_1 \text{ from the estimate}$$

$$|K_0 v|_{L^2(R_y^3)} \leq c_1 |v|_{L^2(\Omega)}$$

for all $v \in L^2(\Omega)$.

We get a bound for the term $- \varepsilon \int_{t_0}^{t} \int_{\Omega} (w_z)^2$ in (4.1) by means of Poincaré's inequality

$$(4.3) \qquad \left(\frac{\alpha}{d}\right)^2 \int_{\mathcal{K}} (v)^2 dz \leq \int_{\mathcal{K}} (v_z)^2 dz \qquad \text{with a constant } \alpha > 0.$$

It holds for all functions $v \in L^2(\mathcal{K})$ with weak derivatives $v_z \in L^2(\mathcal{K})$ and vanishing generalized boundary values on $\partial\mathcal{K}$. Owing to assumption 4.1. (and Fubini's theorem), the function $w(t,y,\cdot)$ belongs to this class for almost all (t,y). By integration of (4.3) with respect to t and y, we get

$$\left(\frac{\alpha}{d}\right)^2 \int_{t_0}^{t} \int_{\Omega} (w)^2 \leq \int_{t_0}^{t} \int_{\Omega} (w_z)^2 ,$$

and from this and (4.1) the integral inequality

$$\varphi(t) \leq \varphi(t_0) + \left\{ c_2 - \varepsilon\left(\frac{\alpha}{d}\right)^2 \right\} \int_{t_0}^{t} \varphi(t')dt'$$

follows for all $o \leq t_0 < t$. Therefore, we can conclude,[5] that the estimate

$$\varphi(t) \leq \varphi(o)\exp(\{ c_2 - \varepsilon(\tfrac{\alpha}{d})^2 \} t)$$

holds for all $t \geq o$. Our result [6] is the

Theorem 4.1.: With the assumption 4.1., stability holds in the case $\frac{c_1 |u^2|_\infty d}{\varepsilon} < \alpha$

[5] c.p. [9, p. 69]
[6] c.p. [8, p. 87] in the case of the Navier-Stokes equation.

References:

1 HANNOSCHÖCK, G., Existenz und Eindeutigkeit bei der Fokker-Planck-Gleichung mit modifiziertem Vlasov-Term, unpublished.

2 HOPF, E.,Über die Anfangswertaufgabe für die hydrodynamischen Grundgleichungen, Math. Nachr. 4 (1951), 213-231.

3 LADYŽENSKAJA, O.A., SOLONNIKOV, V.A., URAL'CEVA, N.N.: Linear and Quasilinear Equations of Parabolic Type, American Mathematical Society, Providence, Rhode Island (1968).

4 RAUTMANN, R., Bemerkungen zur Anfangswertaufgabe einer stabilisierten Navier-Stokesschen Gleichung, ZAMM 55 (1975), T 217-221.

5 RAUTMANN, R., On the Convergence of a Galerkin Method to Solve the Initial Value Problem of a Stabilized Navier-Stokes Equation, ISNM 27 Birkhäuser Verlag, Basel, Stuttgart (1975), 255-264.

6 RAUTMANN, R., Ein konvergentes Hopf-Galerkin-Verfahren für eine Gleichung vom FOKKER-PLANCK-Typ, ZAMM 57 (1977), T 252-253.

7 RAUTMANN, R., The Existence of Weak Solutions of the Fokker-Planck-Vlasov-Equation, to appear in: Methoden und Verfahren d. Math. Physik.

8 SERRIN, J., The Initial Value Problem for the Navier-Stokes Equations, in: Nonlinear Problems (ed. R.E. Langer) MRC Madison (1963), 69-98.

9 WALTER, W. : Differential and Integral Inequalities, Springer Berlin (197o)

1o WAX, N. (ed.) Selected Papers on Noise and Stochastic Processes,Dover, Publ., Inc. New York (1954).

This work has been supported by Forschungsförderung des Landes Nordrhein-Westfalen.

ON ITERATIVE SOLUTION METHODS FOR SYSTEMS OF PARTIAL DIFFERENTIAL EQUATIONS

H.J. WIRZ

I. INTRODUCTION

The solution of coupled systems of "stationary" partial differential equations (p.d.e.'s) by numerical methods almost always requires efficient iterative methods.

It is a well-known fact, that these methods can be thought of solving numerically a continuous artificial or natural evolution problem, which asymptotically should give the desired solution. Most of the presently existing methods are based on "parabolic" p.d.e.'s, while for the large class of hyperbolic equations nothing really is known. The reason for this seems to be obvious, since natural hyperbolic evolution problems are not expected to have an asymptotic stationary solution. The extreme slow convergence of accurate discrete methods to resolve a natural time dependent hyperbolic system then is of course a consequence of small numerical effects (i.e., numerical viscosity).
In order to be independent of the choice of a particular discrete method (Finite Difference, Finite Elements, etc.), we shall consider continuous natural or artificial systems of evolution equations, in particular, hyperbolic p.d.e.'s. Results are reported for a fairly large class of hyperbolic evolution problems, which are called "relaxation" equations. The rate of convergence for these hyperbolic evolution relaxation methods is several orders of magnitude better than for existing dissipative methods for hyperbolic evolution systems without relaxation. See also [1].

II. DIFFUSION AND RELAXATION

Rather than attempting a review of existing methods, we discuss here some aspects of two well-known iterative methods, namely, the ADI (Alternating Direction Implicit) process [2] and the older, but still important, SOR (Successive Overrelaxation)[3,4] method.

Our purpose here is to give some new interpretations for the underlying continuous evolution equations. In order to keep the inevitabl notations as simple as possible, we consider the simplest two-point boundary value problem.

Let $\Omega = \{x | 0 \leq x \leq 1\}$ be a domain with $x \in R_1$. We are looking for a scalar function $\Phi(x)$, satisfying the stationary boundary value problem :

$$\Phi_{xx} = g(x) \quad , \qquad x \in \Omega \tag{1}$$

with homogeneous boundary conditions $\Phi(0) = \Phi(1) = 0$, where $g(x)$ is some given function.

It is not difficult to show, for an infinitely dense grid, that that ADI method leads to the following parabolic evolution problem, in which the independent evolution variable, t, may be called the "time" :

$$\Phi_t - \gamma\Phi_{xx} = g(x) \quad , \qquad t > 0 \quad , \qquad x \in \Omega \quad . \tag{2}$$

The linear occurring constant $\gamma > 0$, is often called the "viscosity" coefficient; it has the physical dimensions of a unit length squared divided by a unit time. The equation (2) is the classic prototype for parabolic p.d.e.'s, and it is not difficult to show, under rather general conditions that any perturbation of $\Phi(x,t)$ will be "damped", the dissipative mechanism being of course "diffusion".

We next consider the SOR process, of which its asymptotic form (infinitely dense grid), [5], for our trivial problem now is a hyperbolic p.d.e., namely

$$\Phi_{tt} + 2k\Phi_t - 2\Phi_{xx} = -2g(x) \quad ; \qquad t > 0 \quad ; \qquad x \in \Omega \quad , \tag{3}$$

where the coefficient k is related to the discrete overrelaxation factor ω_n by $k = 2(2-\omega_n)/\omega_n \Delta X$, in which ΔX denotes the mesh-size. For k to be finite for $\Delta X \to 0$, we have the classical result : $\omega_n = 2-\omega\Delta X/2$, with $\omega > 0$. Note that k is a nonlinear function of ω_n.

The coefficient $\omega > 0$ will be called the relaxation parameter of the continuous evolution problem (3), its inverse the relaxation time τ .

Again, it is not difficult to demonstrate that initially existing perturbations of $\Phi(x,t)$ will be damped out but it is of importance to recognize the different damping mechanism involved. To see this more clearly in physical terms, consider the "natural" wave propagation problem with "damping" :

$$\Phi_{tt} + \frac{1}{\tau}\Phi_t - a^2\Phi_{xx} = -a^2 g(x) \ , \qquad t > 0 \ , \quad x \in \Omega \qquad (4)$$

where Φ may denote a velocity potential, a the velocity of sound. For $\tau \to \infty$, the prototype of a second order hyperbolic p.d.e. is obtained. In contrast to the "diffusion" damping mechanism of parabolic p.d.e.'s, we shall call this dissipative effect "relaxation", in agreement with the physical notations.

Continuing, we rewrite the wave propagation problem with relaxation as a system of evolution equations :

$$\Psi_t - \tau a^2 \Phi_{xx} = -\tau a^2 g(x) \ , \qquad t > 0 \ , \quad x \in \Omega \qquad (5a)$$

$$\Phi_t + \frac{1}{\tau}(\Phi - \Psi) = 0 \ , \qquad (5b)$$

where $\Psi(x,t)$ is another dependent variable. It is not hard to show that this problem is well posed in the usual L_2 sense. The first of these equations resembles very much the parabolic heat conduction equation, to which it reduces, if for $\tau \to \infty$, $\nu = \tau a^2$ remains bounded. We note here only that this property is not just a matter of luck.

The discrete version of equ. (5b) is simply an algebraic relation, an often employed numerical "trick" to accelerate the convergence. There was no explanation, however.

Finally, we consider still another way of representing the hyperbolic evolution problem (4).
Let Φ be the velocity potential, the gradients then are the velocities, and we may write the first order system, defining $u = \Phi_x$, the velocity, and the relaxation variable $\rho = -\Phi_t/a$ to give :

$$u_t + a\rho_x = 0 \ , \qquad t > 0, \quad x \in \Omega. \qquad (6a)$$

$$\rho_t + au_x + \frac{1}{\tau}\rho = ag(x) \ . \qquad (6b)$$

Here the initial values and boundary conditions have to be added. Again, it is not difficult to establish the well-posedness of the problem above (at least for the Cauchy problem).
The above first order hyperbolic evolution problem with relaxation, in which ρ/τ will be called the "relaxation function", represents now the type of new evolution problems we are looking for.

The main problem will be to demonstrate the existence of asymptotic stationary solutions for t tending to infinity, independent of the initial data.

III. SYSTEMS OF EQUATIONS

Let $x = \{x_1, x_2, \ldots, x_r\}$ be a point in a bounded domain Ω with volume V and sufficient smooth surface $\partial\Omega$ with normal (pointing outwards) $n = \{n_1, n_2, \ldots, n_r\}$ of the r-dimensional Euclidian space R_r.

There is a m-vector function $W(x) = \{W_1(x), W_2(x), \ldots, W(x)_m\}$, which denotes the dependent variables. We further introduce the following metric :

$$||u|| = (u,u)^{1/2} = \{\int_V u'u dV\}^{1/2} \tag{7}$$

which defines a Hilbert space over R_r or $C(R_r)$.

3.1 Stationary systems of equations

The stationary systems, in which we are interested in, may generally be formulated as an operator equation of the form

$$P(x,\partial_x)\, W(x) = 0 \tag{8}$$

where $P(x,\partial_x)$ is a general differential operator. We shall assume throughout that the boundary conditions on $\partial\Omega$ are such that a unique stationary solution $W(x)$ may exist.

The general operator $P(x,\partial_x)$ may be split in a sum of differential operators $P_\nu(x,\partial_x)$ of different order

$$P(x,\partial_x) = \sum_{\nu=1}^{s} P_\nu(x,\partial_x) \ , \tag{9}$$

of which we shall consider specifically the following linear, first order operator ($\partial_j = \partial/\partial x_j$) :

$$P_1(x,\partial_x) = \sum_{j=1}^{r} A_j(x)\partial_j \tag{10}$$

where the m×m matrices $A_j(x)$, ($j = 1,2,\ldots,r$) are assumed to be real and symmetric ($A_j = A_j'$), and the linear, second order operator :

$$P_2(x,\partial_x) = - \sum_{j=1}^{r} D_j(x)\partial_j^2 \tag{11}$$

where the real m×m matrices $D_j(x)$, ($j = 1,2,\ldots r$) satisfy an inequality $D_j \geqslant 0$. With these operators (P_1, P_2) most of the practical problems, in particular in fluid dynamics, can be covered.

3.2 Evolution equations

The problem can be formulated as follows : we consider the (in general nonlinear) initial-boundary value problem for a m-vector function $w(x,t)$:

$$w_t + P(x,\partial_x)w(x,t) = 0\,, \qquad t > 0\,,\quad x \in \Omega$$
$$w(x,0) = w_0(x) \qquad t = 0\,. \tag{12}$$

Then the following results are known [5] : suppose there are functions v, $w \in D \subset H$, where D denotes the domain of P, satisfying the boundary conditions and

$$(Pv-Pw,v-w) \geq \gamma\,||v-w||^2; \qquad \gamma > 0\quad . \tag{13}$$

Then for all initial elements u_0, w_0 for which a global solution exists, an estimate can be derived :

$$||v(x,t)-w(x,t)|| \leq ||v_0(x)-w_0(x)||\,\exp(-\gamma t)\ . \tag{14}$$

Thus for t tending to infinity, the solutions are independent of the initial elements. Operators with the property (13) are generally called "dissipative". More specifically, we consider the following evolution problem

$$w_t + P_1(x,\partial_x)w + P_2(x,\partial_x)w = 0\,, \qquad t > 0\,,\quad x \in \Omega$$
$$w(x,0) = w_0(x)\quad, \tag{15}$$

where the operators P_1, P_2 are given by equations 10, 11.

Introducing the transient errors $v(x,t) = w(x,t)-W(x)$, we obtain the corresponding homogeneous problem

$$v_t + P_1 v + P_2 v = 0\,, \qquad t > 0\,,\quad x\ \ \Omega$$
$$v(x,0) = v_0(x) \qquad t = 0 \tag{15a}$$

It is not difficult to show that the following "generalized energy" equation can be derived for all sufficient smooth $v(x,t) \in C^2(R_r)$:

$$\frac{d}{dt}||v||^2 - 2(v,G,v) + 2\sum_{j=1}^{r}(\partial_j v',\tilde{D}_j\partial_j v) + 2S = 0 \tag{16}$$

where the surface integral S is defined as

$$S = \sum_{j=1}^{r}\oint v'\left(\frac{1}{2}A_j v - D_j\partial_j v\right) n_j \qquad \text{and}$$

$$G(x) = \frac{1}{2}\sum_{j=1}^{r}\partial_j A_j\ ;\quad \tilde{D}_j(x) = \frac{1}{2}(D_j + D_j')\ ,\quad (j = 1,2,\ldots,r).$$

3.3 Diffusion and relaxation

(i) Diffusion equations

We start with our evolution problem (15a) for the transient

errors and assume homogeneous (Dirichlet) boundary conditions on $\partial\Omega$, which implies S = 0. For simplicity we further assume constant matrices A_j (j = 1,2,...,r). Then the "generalized energy" equation (16) gives (whatever the matrices A_j are) :

$$\frac{d}{dt} ||v(x,t)||^2 + 2 \sum_{j=1}^{r} (\partial_j v', \tilde{D}_j v) = 0 \tag{17}$$

Thus we see that the "generalized energy" is decreasing in time, provided that the second term is positive for all $v \neq 0$. The damping mechanism involved is, of course, diffusion.

Suppose now that an inequality of the form

$$\sum_{j=1}^{r} (\partial_j v', \tilde{D}_j \, _j v) \geq \gamma ||v||^2 , \qquad \gamma > 0, \quad \text{constant}, \tag{18}$$

can be found[1], then it follows immediately

$$||v(x,t)|| = ||v(x,0)|| \exp(-\gamma t) \tag{19}$$

which implies that the asymptotic solution of our "diffusion" evolution problem is unique and independent of the initial data. Thus we find for large t the unique stationary solution.

In order to get an idea about the rate of convergence, i.e., the number of steps (N) to achieve a certain convergence level for a fixe time, say t^*, we may loosely define

$$N = - \frac{1}{\gamma \Delta t} \ln \frac{||v(x,t^*)||}{||v(x,0)||} \tag{20}$$

in which Δt denotes the time step. Then for explicit methods, where a stability limit $\Delta t \leq s(\Delta x)^2$, $s > 0$, Δx denotes the mesh size, typically holds, we get $N \sim (\Delta x)^{-2}$. The number of "time steps" (iterations) to achieve a certain given convergence level will be proportional to the square of the number of grid points being used. This result can only be improved if there is no stability limit, which is just the case for the ADI methods, for example.

(ii) Relaxation equations

We turn now to hyperbolic evolutions equations and consider the symmetric system of equations for the transient errors v(x,t) togethe with homogeneous boundary conditions.

[1] This is typically the case, if $\tilde{D}_j \geq \delta$, $\delta > 0$, constant (j = 1,2,...,r).

$$v_t + \sum_{j=1}^{r} A_j \partial_j v = 0 \ , \qquad t > 0 \ , \quad x \in \Omega$$
$$v(x,0) = v_0(x) \ , \qquad t = 0 \qquad (21)$$

According to the physics, we introduce the following definitions concerning the contour integral S in equ. (16) :
(i) the boundary conditions are "non-dissipative", if and only if $S = 0$ (the term "neutral" is not very appropriate);
(ii) they are called "dissipative" , if $S > 0$ for all $v \neq 0$.
It is therefore an interesting idea to develop "boundary iterative methods for hyperbolic systems using (ii).

Suppose now that the homogeneous boundary conditions are "non-dissipative" and that the matrices A_j ($j = 1,2,\ldots,r$) are constants, for simplicity. Then we obtain from (16) immediately the differential equation

$$\frac{d}{dt} ||v(x,t)|| = 0 \qquad (22)$$

which implies the conservation of the "generalized energy". It is this fact, which makes it impossible to find asymptotically the (existing) stationary solution.

There are, nevertheless, many time dependent numerical methods available, which for a sufficient large number of time steps do give the desired stationary solution. The contradiction is, of course, easily explained since these methods typically introduce (by the discretization process) a damping mechanism, which is almost always "diffusion". This implies that essentially an evolution system with higher order differential operators is solved in time. For reasons of approximation, the decay constant γ (equ. 18), however, must be a function of the mesh size, ΔX, such that γ vanishes if the mesh size tends to zero. This has an important influence on the rate of convergence.

Suppose for example that the employed explicit method is dissipative of order four (the Lax-Wendroff-scheme [7], for example). Then the decay constant γ is proportional to $(\Delta x)^3$ and taking into account the stability (C.F.L.) limit $\Delta t \leq c\Delta x$, $c > 0$, const., we get from equ. (20) that the number of time steps N is proportional to M^4, where M denotes the number of grid points. The required number of time steps are four orders of magnitude greater than for the simple SOR process. The best one can hope to get with explicit methods is $N \sim M^2$, which is not acceptable for many practical problems.

In order to improve this situation but maintaining the numerous advantages of solving asymptotically stationary first order systems of equations (3-D problems, mixed or hybrid stationary operators) with hyperbolic evolution equations, we consider hyperbolic evolution equations with more dependent variables than the original problem would require.

Typically we consider then systems of evolution equations for a $(m+\ell)$-vector function $v(x,t)$ (the transient errors) of the form ($\ell \geq 1$ is an integer) :

$$v_t + \sum_{j=1}^{r} B_j \partial_j v + \omega L v = 0 , \qquad t > 0 , \quad x \in \Omega \tag{23}$$

$$v(x,0) = v_0(x) , \qquad t = 0.$$

in which the real $(m+\ell)\times(m+\ell)$ matrices B_j ($j = 1,2,...,r$) and L (still to be defined) are given. We shall call the matrix L the "relaxation" matrix, the constant parameter $\omega > 0$, the relaxation coefficient, while the additionally introduced ℓ dependent variables are denoted as relaxation variables. The SOR process, for example, interpreted with the system (equ. 6), has the above structure with $\ell = 1$.

Suppose now that for all $v \in C^1(R_r)$, satisfying the homogeneous, non-dissipative, boundary conditions, an inequality of the form holds :

$$\frac{1}{2} (v',(L+L')v) \geq \gamma \; ||v||^2 , \qquad \gamma > 0 \text{, constant} \tag{24}$$

Then for symmetric matrices B_j ($j = 1,2,...,r$) with constant coefficients an estimate for the generalized energy for the hyperbolic evolution problem can easily be derived :

$$||v(x,t)|| \leq || v(x,0) || \exp(-\omega\gamma t) . \tag{25}$$

Using the same arguments as before, and assuming the existence of a stationary solution, this solution will be found uniquely for sufficient large times, independent of the initial data. Thus the hyperbolic evolution system (equ. 23) is representing a "relaxation process" with a damping mechanism essentially different from that of "diffusion".

There remains to specify the matrices B_j ($j = 1,2,...,r$) and the relaxation matrix L. For simplicity, we consider first the class of symmetric hyperbolic evolution problems (as they occur in fluid dynamics, for example).

For these problems, it is sufficient to introduce a single additional relaxation variable, denoted as $\xi(x,t)$. Defining the two m-vector functions, where a,b are some constant vectors :

$$p = w(x,t) + a\xi(x,t) \qquad \text{and} \qquad q = w(x,t) + b\xi(x,t) \tag{26}$$

we alter the original symmetric hyperbolic evolution problem :

$$w_t + \sum_{j=1}^{r} A_j \partial_j w = 0$$

such that a symmetric relaxation evolution system is obtained :

$$\begin{aligned} & w_t + \sum_{j=1}^{r} A_j \partial_j p = 0 \; ; \qquad t > 0 \; , \quad x \in \Omega \\ & \xi_t + a' \sum_{j=1}^{r} A_j \partial_j q + \omega \xi = 0 \end{aligned} \qquad (27)$$

with initial conditions $w(x,0) = w_0(x)$; $\xi(x,0) = \xi_0(x)$. Introducing the $(m+1)$-vector function $u(x,t) = \{w,\xi\}$, the above system is of the form (23), where the structure of the matrices B_j and L are given below :

$$u = \begin{pmatrix} w \\ \xi \end{pmatrix} \; ; \qquad B_j = \left(\begin{array}{c|c} A_j & A_j a \\ \hline a' A_j & a' A_j b \end{array}\right) \; ; \qquad L = \left(\begin{array}{c|c} 0 & 0 \\ \hline 0 & 1 \end{array}\right) .$$

It remains to demonstrate that for this system an estimate of the type of equation (24) holds. This turns out to be more difficult, since the relaxation matrix is only positive semi-definit. Without going into technical details, the essential condition is the following :
Suppose that each eigenvalue of the matrix $\sum_{j=1}^{r} B_j k_j$ with $\sum_{j=1}^{r} k_j^2 = 1$ and $|k_j| \leq 1$, k_j real $(j = 1,2,\ldots,r)$, is different from each eigenvalue of the matrix $\sum_{j=1}^{r} A_j k_j$, then for every fixed relaxation coefficient $\omega > 0$ and with $|a| \neq 0$; $|b| \neq 0$, there exists a positive decay constant $\gamma < 1$ (equ.24). Thus, as with any other acceleration method, a shift of the locus of the eigenvalues is indispensable.

Furthermore, it is not difficult to show that the relaxation system (27) indeed has this property.

In order to estimate the rate of convergence we again consider the number of time steps (iterations), N, as given by the formula (20). Since the leading term of the decay constant, γ, of the (discrete) hyperbolic relaxation system (27) is independent of the mesh size, Δx, and taking into account a typical stability constraint (C.F.L.) for explicit methods $\Delta t \leq c\Delta x$, $c > 0$, we conclude that the number N is essentially proportional to $(\Delta x)^{-1}$. Thus N is of the same order as the SOR process. Remembering that for the Lax-Wendroff method, for example, the number N is about three orders of magnitude greater, we see that the above hyperbolic relaxation process represents a significant reduction of the computation time.

Before considering some examples; we conclude with some remarks.
The question arises, whether the matrices A_j ($j = 1,2,\ldots,r$) have to be symmetric (or more general, Hermitian). There is evidence that the ideas above can be applied for strictly hyperbolic systems, which are not symmetric.
There exists, however, always the possibility to increase the number, (ℓ), of the relaxation variables (at the expense of more storage requirements), to convert any hyperbolic system into a relaxation process.
Finally, the independent variables, x, need not to be spatial coordinates the evolution variable needs not to be the time. This offers a number of new relaxation methods for certain degenerated physical time dependent problems.

IV. SOME EXAMPLES

The following few examples are taken from problems in fluid mechanics, (see also [1]). For simplicity we consider stationary problems in two independent variables, denoted as x,y.

4.1 Inviscid, irrotational and rotational flows

The stationary equations are essentially :

$$\begin{aligned} &u_x + v_y = 0 \\ &-v_x + u_y = \rho(x,y) \end{aligned} \qquad (28)$$

in which u, v are the velocity components and $\rho(x,y)$ the given vorticity As an example of a relaxation process for the above system we may consider :

$$\begin{aligned} &u_t + u_x + v_y + \lambda\xi_x = 0 \\ &v_t - v_x + u_y + \lambda\xi_y = \rho(x,y) \\ &\xi_t + \lambda(u_x+v_y) + \lambda^2\xi_x + \omega\xi = 0 \end{aligned} \qquad (28a)$$

in which ξ denotes the relaxation variable, λ a constant and $\omega > 0$ the inverse of the relaxation time.

4.2 Incompressible, inviscid and viscous flows

The stationary system is the following, expressing the conservation of mass and momentum (p denotes the "pressure", Re denotes the Reynolds number) :

$$\begin{aligned} &u_x + v_y = 0 ; \\ &uu_x + vu_y + p_x - \frac{1}{Re}\Delta u = 0 ; \qquad \Delta = \frac{\partial^2}{\partial x^2} + \frac{\partial^2}{\partial y^2} \\ &uv_x + vv_y + p_y - \frac{1}{Re}\Delta v = 0 ; \end{aligned} \qquad (29)$$

The relaxation process for this (quasilinear) problem reads,

$$\begin{aligned} &p_t + u_x + v_y = 0 \\ &u_t + uu_x + vu_y + p_x + \lambda\xi_x - \frac{1}{Re}\Delta u = 0 \\ &v_t + uv_x + vv_y + p_y + \lambda\xi_y - \frac{1}{Re}\Delta v = 0 \\ &\xi_t + \lambda(u_x+v_y) + \omega\xi = 0 \end{aligned} \qquad (29a)$$

The above relaxation process is extremely difficult to analyze, since the problem is now non-linear. "Freezing" the coefficients, this evolution problem belongs to the class considered earlier.

4.3 Irrotational, inviscid, transonic flows

The final example will be a case where the stationary operator is nonlinear and changes the type : for subsonic flows the operator is elliptic, it is hyperbolic for supersonic flows. For simplicity, we consider the "small disturbances" equations :

$$\begin{aligned} & f_x(u) + v_y = 0 \qquad\qquad f(u) = (1-M_\infty^2)u + \frac{\kappa-1}{2} M_\infty^2 u^2 \\ & -v_x + u_y = 0 \end{aligned} \tag{30}$$

in which M_∞ denotes the constant free stream Mach number and κ the ratio of specific heats (1.4 for air, for example).
A simple relaxation process for the above system, maintaining the conservation property (shocks!), then is :

$$\begin{aligned} & u_t + f_x(u+\lambda\xi) + v_y = 0 \\ & v_t - v_x + (u+\lambda\xi)_y = 0 \\ & \xi_t + \lambda f_x(u+\mu\xi) + \lambda v_y + \omega\xi = 0 \end{aligned} \tag{31}$$

in which λ, μ and $\omega > 0$ are constants. The constants λ, μ have to be selected such that an entropy condition in the sense of Lax [8] is met. The last (discrete) relaxation system offers, apart from the interesting rate of convergence, another important numerical advantage : Most of the present methods to treat stationary transonic flows with the so-called "shock capturing" technique require a complicated type dependent differencing of the stationary operator. This is completely avoided here since any of the existing methods for time dependent hyperbolic conservation equations will do. The problem of optimization, however, is not easy to solve.

5. REFERENCES

1. WIRZ, H.J.: Relaxation methods for time dependent conservation equations in fluid mechanics.
AGARD LS 86, 1977.
2. PEACEMAN, D.W. & RACHFORD, H.H: The numerical solution of parabolic and elliptic differential equations.
SIAM 3, 1955, pp 28-41.
3. FRANKEL, S.P.: Convergence rates of iterative treatments of partial differential equations.
MTCA, Vol. 4, 1950, pp 65-75.
4. YOUNG, D.: Iterative methods for solving partial differential equations of elliptic type.
Am. Math. Soc. Transact., Vol. 75, 1954, pp 92-111.
5. GARABEDIAN, P.: Estimation of the relaxation factor for small mesh size.
Math. Tables Aids Comp. Vol. 10, 1956, pp 183-185.
6. VAINBERG, M.M.: Variational method and method of monotone operators in the theory of nonlinear equations.
John Wiley, New York, 1973.
7. LAX, P.D. & WENDROFF, B.: Difference schemes with high order of accuracy for solving hyperbolic equations.
Comm. Pure & Appl.Math., Vol. 17, 1964, pp 381.
8. LAX, P.D.: Hyperbolic systems of conservation laws and the mathematical theory of shock waves.
SIAM, Philadelphia, 1973.

REFERENCES

1. [illegible] for [illegible] conservation equations in fluid mechanics [illegible]
2. [illegible] The numerical solution of [illegible] differential equation [illegible]
3. [illegible] convergence [illegible] partial differential [illegible]
4. [illegible] equations [illegible]
5. [illegible] the [illegible] factor for [illegible] mean [illegible]
6. [illegible] method [illegible] equations [illegible]
7. [illegible] difference [illegible] with high [illegible] hyperbolic equations [illegible] Math. [illegible]
8. [illegible] systems [illegible] and [illegible]

'ol. 521: G. Cherlin, Model Theoretic Algebra - Selected Topics. V, 234 pages. 1976.

'ol. 522: C. O. Bloom and N. D. Kazarinoff, Short Wave Radiation Problems in Inhomogeneous Media: Asymptotic Solutions. V. 104 pages. 1976.

'ol. 523: S. A. Albeverio and R. J. Høegh-Krohn, Mathematical Theory of Feynman Path Integrals. IV, 139 pages. 1976.

'ol. 524: Séminaire Pierre Lelong (Analyse) Année 1974/75. Edité par P. Lelong. V, 222 pages. 1976.

'ol. 525: Structural Stability, the Theory of Catastrophes, and Applications in the Sciences. Proceedings 1975. Edited by P. Hilton. VI, 408 pages. 1976.

'ol. 526: Probability in Banach Spaces. Proceedings 1975. Edited by A. Beck. VI, 290 pages. 1976.

'ol. 527: M. Denker, Ch. Grillenberger, and K. Sigmund, Ergodic Theory on Compact Spaces. IV, 360 pages. 1976.

'ol. 528: J. E. Humphreys, Ordinary and Modular Representations of Chevalley Groups. III, 127 pages. 1976.

'ol. 529: J. Grandell, Doubly Stochastic Poisson Processes. X, 234 pages. 1976.

'ol. 530: S. S. Gelbart, Weil's Representation and the Spectrum of the Metaplectic Group. VII, 140 pages. 1976.

'ol. 531: Y.-C. Wong, The Topology of Uniform Convergence on Order-Bounded Sets. VI, 163 pages. 1976.

'ol. 532: Théorie Ergodique. Proceedings 1973/1974. Edité par J.-P. Conze and M. S. Keane. VIII, 227 pages. 1976.

'ol. 533: F. R. Cohen, T. J. Lada, and J. P. May, The Homology of Iterated Loop Spaces. IX, 490 pages. 1976.

'ol. 534: C. Preston, Random Fields. V, 200 pages. 1976.

'ol. 535: Singularités d'Applications Differentiables. Plans-sur-Bex. 1975. Edité par O. Burlet et F. Ronga. V, 253 pages. 1976.

'ol. 536: W. M. Schmidt, Equations over Finite Fields. An Elementary Approach. IX, 267 pages. 1976.

'ol. 537: Set Theory and Hierarchy Theory. Bierutowice, Poland 1975. A Memorial Tribute to Andrzej Mostowski. Edited by W. Marek, M. Srebrny and A. Zarach. XIII, 345 pages. 1976.

'ol. 538: G. Fischer, Complex Analytic Geometry. VII, 201 pages. 1976.

'ol. 539: A. Badrikian, J. F. C. Kingman et J. Kuelbs, Ecole d'Eté de Probabilités de Saint Flour V-1975. Edité par P.-L. Hennequin. IX, 314 pages. 1976.

'ol. 540: Categorical Topology, Proceedings 1975. Edited by E. Binz and H. Herrlich. XV, 719 pages. 1976.

'ol. 541: Measure Theory, Oberwolfach 1975. Proceedings. Edited by A. Bellow and D. Kölzow. XIV, 430 pages. 1976.

'ol. 542: D. A. Edwards and H. M. Hastings, Čech and Steenrod Homotopy Theories with Applications to Geometric Topology. VII, 296 pages. 1976.

'ol. 543: Nonlinear Operators and the Calculus of Variations, Bruxelles 1975. Edited by J. P. Gossez, E. J. Lami Dozo, J. Mawhin, and L. Waelbroeck, VII, 237 pages. 1976.

'ol. 544: Robert P. Langlands, On the Functional Equations Satisfied by Eisenstein Series. VII, 337 pages. 1976.

'ol. 545: Noncommutative Ring Theory. Kent State 1975. Edited by J. H. Cozzens and F. L. Sandomierski. V, 212 pages. 1976.

'ol. 546: K. Mahler, Lectures on Transcendental Numbers. Edited and Completed by B. Diviš and W. J. Le Veque. XXI, 254 pages. 1976.

'ol. 547: A. Mukherjea and N. A. Tserpes, Measures on Topological Semigroups: Convolution Products and Random Walks. V, 197 pages. 1976.

'ol. 548: D. A. Hejhal, The Selberg Trace Formula for PSL (2, ℝ). Volume I. VI, 516 pages. 1976.

'ol. 549: Brauer Groups, Evanston 1975. Proceedings. Edited by D. Zelinsky. V, 187 pages. 1976.

'ol. 550: Proceedings of the Third Japan – USSR Symposium on Probability Theory. Edited by G. Maruyama and J. V. Prokhorov. VI, 722 pages. 1976.

Vol. 551: Algebraic K-Theory, Evanston 1976. Proceedings. Edited by M. R. Stein. XI, 409 pages. 1976.

Vol. 552: C. G. Gibson, K. Wirthmüller, A. A. du Plessis and E. J. N. Looijenga. Topological Stability of Smooth Mappings. V, 155 pages. 1976.

Vol. 553: M. Petrich, Categories of Algebraic Systems. Vector and Projective Spaces, Semigroups, Rings and Lattices. VIII, 217 pages. 1976.

Vol. 554: J. D. H. Smith, Mal'cev Varieties. VIII, 158 pages. 1976.

Vol. 555: M. Ishida, The Genus Fields of Algebraic Number Fields. VII, 116 pages. 1976.

Vol. 556: Approximation Theory. Bonn 1976. Proceedings. Edited by R. Schaback and K. Scherer. VII, 466 pages. 1976.

Vol. 557: W. Iberkleid and T. Petrie, Smooth S^1 Manifolds. III, 163 pages. 1976.

Vol. 558: B. Weisfeiler, On Construction and Identification of Graphs. XIV, 237 pages. 1976.

Vol. 559: J.-P. Caubet, Le Mouvement Brownien Relativiste. IX, 212 pages. 1976.

Vol. 560: Combinatorial Mathematics, IV, Proceedings 1975. Edited by L. R. A. Casse and W. D. Wallis. VII, 249 pages. 1976.

Vol. 561: Function Theoretic Methods for Partial Differential Equations. Darmstadt 1976. Proceedings. Edited by V. E. Meister, N. Weck and W. L. Wendland. XVIII, 520 pages. 1976.

Vol. 562: R. W. Goodman, Nilpotent Lie Groups: Structure and Applications to Analysis. X, 210 pages. 1976.

Vol. 563: Séminaire de Théorie du Potentiel. Paris, No. 2. Proceedings 1975–1976. Edited by F. Hirsch and G. Mokobodzki. VI, 292 pages. 1976.

Vol. 564: Ordinary and Partial Differential Equations, Dundee 1976. Proceedings. Edited by W. N. Everitt and B. D. Sleeman. XVIII, 551 pages. 1976.

Vol. 565: Turbulence and Navier Stokes Equations. Proceedings 1975. Edited by R. Temam. IX, 194 pages. 1976.

Vol. 566: Empirical Distributions and Processes. Oberwolfach 1976. Proceedings. Edited by P. Gaenssler and P. Révész. VII, 146 pages. 1976.

Vol. 567: Séminaire Bourbaki vol. 1975/76. Exposés 471–488. IV, 303 pages. 1977.

Vol. 568: R. E. Gaines and J. L. Mawhin, Coincidence Degree, and Nonlinear Differential Equations. V, 262 pages. 1977.

Vol. 569: Cohomologie Etale SGA 4½. Séminaire de Géométrie Algébrique du Bois-Marie. Edité par P. Deligne. V, 312 pages. 1977.

Vol. 570: Differential Geometrical Methods in Mathematical Physics, Bonn 1975. Proceedings. Edited by K. Bleuler and A. Reetz. VIII, 576 pages. 1977.

Vol. 571: Constructive Theory of Functions of Several Variables, Oberwolfach 1976. Proceedings. Edited by W. Schempp and K. Zeller. VI. 290 pages. 1977

Vol. 572: Sparse Matrix Techniques, Copenhagen 1976. Edited by V. A. Barker. V, 184 pages. 1977.

Vol. 573: Group Theory, Canberra 1975. Proceedings. Edited by R. A. Bryce, J. Cossey and M. F. Newman. VII, 146 pages. 1977.

Vol. 574: J. Moldestad, Computations in Higher Types. IV, 203 pages. 1977.

Vol. 575: K-Theory and Operator Algebras, Athens, Georgia 1975. Edited by B. B. Morrel and I. M. Singer. VI, 191 pages. 1977.

Vol. 576: V. S. Varadarajan, Harmonic Analysis on Real Reductive Groups. VI, 521 pages. 1977.

Vol. 577: J. P. May, E_∞ Ring Spaces and E_∞ Ring Spectra. IV, 268 pages. 1977.

Vol. 578: Séminaire Pierre Lelong (Analyse) Année 1975/76. Edité par P. Lelong. VI, 327 pages. 1977.

Vol. 579: Combinatoire et Représentation du Groupe Symétrique, Strasbourg 1976. Proceedings 1976. Edité par D. Foata. IV, 339 pages. 1977.

Vol. 580: C. Castaing and M. Valadier, Convex Analysis and Measurable Multifunctions. VIII, 278 pages. 1977.

Vol. 581: Séminaire de Probabilités XI, Université de Strasbourg. Proceedings 1975/1976. Edité par C. Dellacherie, P. A. Meyer et M. Weil. VI, 574 pages. 1977.

Vol. 582: J. M. G. Fell, Induced Representations and Banach *-Algebraic Bundles. IV, 349 pages. 1977.

Vol. 583: W. Hirsch, C. C. Pugh and M. Shub, Invariant Manifolds. IV, 149 pages. 1977.

Vol. 584: C. Brezinski, Accélération de la Convergence en Analyse Numérique. IV, 313 pages. 1977.

Vol. 585: T. A. Springer, Invariant Theory. VI, 112 pages. 1977.

Vol. 586: Séminaire d'Algèbre Paul Dubreil, Paris 1975–1976 (29ème Année). Edited by M. P. Malliavin. VI, 188 pages. 1977.

Vol. 587: Non-Commutative Harmonic Analysis. Proceedings 1976. Edited by J. Carmona and M. Vergne. IV, 240 pages. 1977.

Vol. 588: P. Molino, Théorie des G-Structures: Le Problème d'Equivalence. VI, 163 pages. 1977.

Vol. 589: Cohomologie l-adique et Fonctions L. Séminaire de Géométrie Algébrique du Bois-Marie 1965–66, SGA 5. Edité par L. Illusie. XII, 484 pages. 1977.

Vol. 590: H. Matsumoto, Analyse Harmonique dans les Systemes de Tits Bornologiques de Type Affine. IV, 219 pages. 1977.

Vol. 591: G. A. Anderson, Surgery with Coefficients. VIII, 157 pages. 1977.

Vol. 592: D. Voigt, Induzierte Darstellungen in der Theorie der endlichen, algebraischen Gruppen. V, 413 Seiten. 1977.

Vol. 593: K. Barbey and H. König, Abstract Analytic Function Theory and Hardy Algebras. VIII, 260 pages. 1977.

Vol. 594: Singular Perturbations and Boundary Layer Theory, Lyon 1976. Edited by C. M. Brauner, B. Gay, and J. Mathieu. VIII, 539 pages. 1977.

Vol. 595: W. Hazod, Stetige Faltungshalbgruppen von Wahrscheinlichkeitsmaßen und erzeugende Distributionen. XIII, 157 Seiten. 1977.

Vol. 596: K. Deimling, Ordinary Differential Equations in Banach Spaces. VI, 137 pages. 1977.

Vol. 597: Geometry and Topology, Rio de Janeiro, July 1976. Proceedings. Edited by J. Palis and M. do Carmo. VI, 866 pages. 1977.

Vol. 598: J. Hoffmann-Jørgensen, T. M. Liggett et J. Neveu, Ecole d'Eté de Probabilités de Saint-Flour VI – 1976. Edité par P.-L. Hennequin. XII, 447 pages. 1977.

Vol. 599: Complex Analysis, Kentucky 1976. Proceedings. Edited by J. D. Buckholtz and T. J. Suffridge. X, 159 pages. 1977.

Vol. 600: W. Stoll, Value Distribution on Parabolic Spaces. VIII, 216 pages. 1977.

Vol. 601: Modular Functions of oneVariableV, Bonn 1976. Proceedings. Edited by J.-P. Serre and D. B. Zagier. VI, 294 pages. 1977.

Vol. 602: J. P. Brezin, Harmonic Analysis on Compact Solvmanifolds. VIII, 179 pages. 1977.

Vol. 603: B. Moishezon, Complex Surfaces and Connected Sums of Complex Projective Planes. IV, 234 pages. 1977.

Vol. 604: Banach Spaces of Analytic Functions, Kent, Ohio 1976. Proceedings. Edited by J. Baker, C. Cleaver and Joseph Diestel. VI, 141 pages. 1977.

Vol. 605: Sario et al., Classification Theory of Riemannian Manifolds. XX, 498 pages. 1977.

Vol. 606: Mathematical Aspects of Finite Element Methods. Proceedings 1975. Edited by I. Galligani and E. Magenes. VI, 362 pages. 1977.

Vol. 607: M. Métivier, Reelle und Vektorwertige Quasimartingale und die Theorie der Stochastischen Integration. X, 310 Seiten. 1977.

Vol. 608: Bigard et al., Groupes et Anneaux Réticulés. XIV, 334 pages. 1977.

Vol. 609: General Topology and Its Relations to Modern Analysis and Algebra IV. Proceedings 1976. Edited by J. Novák. XVIII, 225 pages. 1977.

Vol. 610: G. Jensen, Higher Order Contact of Submanifolds of Homogeneous Spaces. XII, 154 pages. 1977.

Vol. 611: M. Makkai and G. E. Reyes, First Order Categorical Logic. VIII, 301 pages. 1977.

Vol. 612: E. M. Kleinberg, Infinitary Combinatorics and the Axiom of Determinateness. VIII, 150 pages. 1977.

Vol. 613: E. Behrends et al., L^p-Structure in Real Banach Spaces. X, 108 pages. 1977.

Vol. 614: H. Yanagihara, Theory of Hopf Algebras Attached to Group Schemes. VIII, 308 pages. 1977.

Vol. 615: Turbulence Seminar, Proceedings 1976/77. Edited by P. Bernard and T. Ratiu. VI, 155 pages. 1977.

Vol. 616: Abelian Group Theory, 2nd New Mexico State University Conference, 1976. Proceedings. Edited by D. Arnold, R. Hunter and E. Walker. X, 423 pages. 1977.

Vol. 617: K. J. Devlin, The Axiom of Constructibility: A Guide for the Mathematician. VIII, 96 pages. 1977.

Vol. 618: I. I. Hirschman, Jr. and D. E. Hughes, Extreme Eigen Values of Toeplitz Operators. VI, 145 pages. 1977.

Vol. 619: Set Theory and Hierarchy Theory V, Bierutowice 1976. Edited by A. Lachlan, M. Srebrny, and A. Zarach. VIII, 358 pages. 1977.

Vol. 620: H. Popp, Moduli Theory and Classification Theory of Algebraic Varieties. VIII, 189 pages. 1977.

Vol. 621: Kauffman et al., The Deficiency Index Problem. VI, 112 pages. 1977.

Vol. 622: Combinatorial Mathematics V, Melbourne 1976. Proceedings. Edited by C. Little. VIII, 213 pages. 1977.

Vol. 623: I. Erdelyi and R. Lange, Spectral Decompositions on Banach Spaces. VIII, 122 pages. 1977.

Vol. 624: Y. Guivarc'h et al., Marches Aléatoires sur les Groupes de Lie. VIII, 292 pages. 1977.

Vol. 625: J. P. Alexander et al., Odd Order Group Actions and Witt Classification of Innerproducts. IV, 202 pages. 1977.

Vol. 626: Number Theory Day, New York 1976. Proceedings. Edited by M. B. Nathanson. VI, 241 pages. 1977.

Vol. 627: Modular Functions of One Variable VI, Bonn 1976. Proceedings. Edited by J.-P. Serre and D. B. Zagier. VI, 339 pages. 1977.

Vol. 628: H. J. Baues, Obstruction Theory on the Homotopy Classification of Maps. XII, 387 pages. 1977.

Vol. 629: W. A. Coppel, Dichotomies in Stability Theory. VI, 98 pages. 1978.

Vol. 630: Numerical Analysis, Proceedings, Biennial Conference, Dundee 1977. Edited by G. A. Watson. XII, 199 pages. 1978.

Vol. 631: Numerical Treatment of Differential Equations. Proceedings 1976. Edited by R. Bulirsch, R. D. Grigorieff, and J. Schröder. X, 219 pages. 1978.

Vol. 632: J.-F. Boutot, Schéma de Picard Local. X, 165 pages. 1978.

Vol. 633: N. R. Coleff and M. E. Herrera, Les Courants Résiduels Associés à une Forme Méromorphe. X, 211 pages. 1978.

Vol. 634: H. Kurke et al., Die Approximationseigenschaft lokaler Ringe. IV. 204 Seiten. 1978.

Vol. 635: T. Y. Lam, Serre's Conjecture. XVI, 227 pages. 1978.

Vol. 636: Journées de Statistique des Processus Stochastiques, Grenoble 1977, Proceedings. Edité par Didier Dacunha-Castelle et Bernard Van Cutsem. VII, 202 pages. 1978.

Vol. 637: W. B. Jurkat, Meromorphe Differentialgleichungen. VII, 194 Seiten. 1978.

Vol. 638: P. Shanahan, The Atiyah-Singer Index Theorem, An Introduction. V, 224 pages. 1978.

Vol. 639: N. Adasch et al., Topological Vector Spaces. V, 125 pages. 1978.